Voyage to the Milky Way

The Future of Space Exploration

VOYAGE TO THE MILKY WAY

The Future of Space Exploration

Donald Goldsmith

Publisher's Cataloging-in-Publication Data
Goldsmith, Donald.
Voyage to the Milky Way : the future of space exploration / by Donald Goldsmith. — 1st ed.
p. cm.
Includes bibliographical references and index.
ISBN: 1-57500-046-6

1.Space flight—Forecasting. 2. Interplanetary voyages. I.Title.

L790.G65 1999 629.4
QBI99–453

Book design by Joe Gannon

Manufactured in the United States of America

TV Books. L.L.C.
Publishers serving the television industry.
1619 Broadway, Ninth Floor
New York, NY 10019
www.tvbooks.com

To Rachel on her voyage

Contents

Acknowledgments

I would like to thank my friends and relatives who helped me while I was writing this book: Marty Africa, Bruce and Kathy Armbruster, Sam Bader, Charles Beichman, Arnie Berger, Aviva and Kenneth Brecher, Pascal Debergue, John and Michelle Edelsberg, George and Susan Field, June Fox, Marjorie and Victor Garlin, Jane Goldsmith, Nicholas and Marianne Goldsmith, Paul Goldsmith, Rachel Goldsmith, Amy and George Gorman, Dolly and Don Hatch, Jerry and Marjorie Heymann, Carol Kraus, Don and Linda Kripke, Jon and Sharona Lomberg, Tom Lucas, Sally and Stephen Maran, Ellen and Laurence Marschall, Tom McGuire, David Morrison, George Musser, Craig and Christine Nova, Eleanor and Alexander Orr, Tobias and Natasha Owen, Merrinell Phillips, Astrid and Howard Preston, Jo Powe, Arlene Prunella, Sheryl Reiss, Michael, Kristy, Paul, and Rita Sagalyn, Brian Salzberg, Anneila Sargent, Frank Shu, Peter Stine, and Jerry Wasserburg.

Voyage to the Milky Way

The Future of Space Exploration

Introduction

The second half of the twentieth century marks the time when humanity began to leave its home on Earth to explore the host of other objects in orbit around the sun. This exploration, which has already opened the solar system to human inspection, will eventually take human beings to the sun's other planets, and some day, if we survive, to planetary systems around other stars. Our studies of the solar system with automated probes have led to the discovery of startlingly diverse worlds among nearby celestial objects. From the dusty plains of Mars, in whose frozen soil we may yet find life, to the frozen surface of Jupiter's moon Europa, which may conceal a worldwide ocean; from the ethane lakes on Saturn's moon Titan to the asteroids that may contain a fortune in precious metals, our celestial neighbors conceal myriad treasures and possibilities. As Carl Sagan put it, a person would have to be made of wood not to become excited at the prospect of learning more about these other worlds.

Humans yearn to connect with the universe and to find meaning in the cosmos. We are an exploring species, curious and restless. These attitudes and emotions support our spiritual longings and undergird our efforts to understand nature. They are also linked to a more problematic side of human nature, the desire to appropriate as much of nature's bounty to ourselves as we possibly can. In our earlier evolution, this stance made perfect sense, but it has caused difficulties as we have come to dominate our planet, growing in numbers that ceaselessly demand to be fed and sustained.

To some, the cosmos offers the solution and inspiration for our future development. Rich in minerals and other chemical compounds, unfettered by governmental laws and regulations, the objects in our solar system might soon become a new human frontier, booming with energy and experimentation. Space advocacy enthusiasts go so far as to predict that if we fail to send humans to colonize other worlds, humanity will perish.

When considering the possibilities for colonizing other worlds, and of transforming them so that they acquire more Earth-like conditions, I find myself in the opposing, go-slow camp. While I cannot count myself a full member of the cult of decorous behavior, I firmly believe that we ought to clean up more of Earth before we plant ourselves on other worlds. In matters of exploration, I remain a total enthusiast—but I am convinced that we should devote most of our efforts to sending ever more capable automated explorers, which cost and risk far less than human voyagers do. To those who insist that humans can truly identify only with others of our species, I submit that we can learn to enjoy the results from the space probes we have built as well and as eagerly as we now listen to human astronauts.

Voyage to the Milky Way offers a summary of our future prospects for exploring the vast cosmos in which Earth floats, our blue-green home in the heavens. In creating this summary, I have attempted to balance what we may learn from automated spacecraft with the possibilities for human spaceflight. The creative tension between my own views (surely not unique, of course) and the opinions of those who pant for human missions to the moon and planets has helped me in writing this book and will, I hope, make its pages more vivid for my readers. In any case, I offer you a tour through our solar system and beyond, into the immense realm of the Milky Way galaxy. Ride along and form your own views—or, if you prefer, fall subject to my own. In either case, take a moment to think about the human endeavor to explore the universe, a small part of which has already occurred, with far greater adventures to come.

Berkeley, California
March 1999

Chapter 1

Whither Go We?

The year 1999, whose nines signal the coming of a new millennium, also marks the eightieth anniversary of an obscure research publication, "A Method of Reaching Extreme Altitudes." In that scientific paper, Robert Goddard, an assistant professor of chemistry at Clark University in Worcester, Massachusetts, glimpsed the future of rocketry. From Goddard's basic design for a liquid-fueled rocket came the mighty engines that have sent astronauts to the moon and space probes to all the planets in our solar system but one. Once mocked as "moon-mad" by *The New York Times,* Goddard did not live to see the Goddard Space Flight Center named in his honor. The National Aeronautics and Space Administration (NASA) uses this center as its prime site for the management of automated Earth satellites that have introduced new ways for humanity to view the cosmos. But Goddard's concepts opened the era of space exploration, the epoch in which humanity saw how to extend its reach through space to arrive at other celestial objects.

At a point almost halfway between 1919 and 1999, on October 4, 1957, a race into space gave Earth its first artificial satellite, the Soviet Union's *Sputnik 1.* Forty years ago, at the start of 1959, the Soviet Union and the United States had sent a total of only nine satellites into orbit around Earth, and at least ten other launch attempts had ended in failure. Yet all humanity could see that far greater successes lay in the future; indeed, as 1959 opened, the Soviet Union sent a probe close to the moon—humanity's first emissary to another world—and two years later sent the first astronauts into orbit around Earth. Thirty years ago, on July 20, 1969, the United States's furiously energetic lunar landing program reached a culmination as Neil Armstrong became the first human to walk on another celestial object.

During the next three decades, space exploration entered its adolescence as a host of countries participated in creating a vast array of satellites and spacecraft capable of myriad tasks for extending our knowledge of both our own planet and its cosmic surroundings and to improve our ability to communicate with one another. Thousands of separate objects, designed and fabricated by more than three dozen countries, have now orbited Earth. Four hundred astronauts have spent a combined total of nearly ten thousand days in space, with the individual record (nearly one year and three months) set by the Russian Valerii Polyakov in 1994–95. Twelve men have walked or ridden on the lunar surface, spending as much as seven hours at a time exploring our neighboring world. We have built and launched more than one hundred probes to make close-up observations of three dozen objects in the solar system, including Mercury, Venus, Mars, Jupiter and its four large moons, Saturn and its system of rings and satellites, Uranus, Neptune, Halley's comet, and the asteroids Gaspra and Ida. Can the stars be far beyond?

The Arrangement of the Cosmos

The twentieth century has carried humanity from dreams of a space age into a complex, ongoing set of contacts with the cosmos, itself revealed by a host of new instruments in space and on Earth. We now know that our solar system forms a tiny part of the Milky Way galaxy, a giant assemblage of stars, shaped like a pie plate with an ice cream bulge at its center. The sun and its gravitationally bound family of planets, moons, asteroids, meteoroids, and comets reside within the outer reaches of the plate-shaped Milky Way, closer to our galaxy's edge than to its center. Though we correctly call the Milky Way "our galaxy," this title belongs to all of the galaxy's three hundred billion stars. In addition to its stars and their planets (the latter discovered only during the past few years), the Milky Way contains enormous amounts of interstellar gas and dust that float among the stars, plus a host of exotic objects to which astronomers have given suggestive names: white dwarfs, red giants, black holes, neutron stars, dark molecular clouds, planetary nebulae, star-forming regions, and supernova remnants.

Every probe that we launch into space travels through the Milky Way, but for only a tiny part of the galaxy's total extent. The Milky Way's diameter covers twenty-five thousand times the distance from the sun to its closest neighboring stars. These closest stars in turn lie twenty-five thousand times farther from us than the farthest planets in our solar system, the most distant objects to which we have sent spacecraft. Thus a spacecraft voyaging from one side of the Milky Way to the other would travel about 625 million times farther than the longest space journeys—measured in billions of miles—that we have yet achieved with automated spacecraft.

Planning for the Future of Space Exploration

Nevertheless, we have begun to explore the Milky Way, starting with our immediate environment and looking outward toward more distant objects, which will require still longer journeys if we hope for close-up inspection. An ideal society would formulate a coherent, long-term plan to explore the cosmos, modifying it as new discoveries and changed circumstances warrant. This plan would lead us from the solar system outward, eventually taking us tens of thousands of times farther than the sun's most distant planets and finally carrying us to the stars at the appropriate time, when humanity has achieved the technological and sociological insight required for such extensive voyages.

Our attitudes toward space exploration well demonstrate that we do not yet possess an ideal society. National governments continually change their attitudes toward the importance of spending money on cosmic research, especially and understandably during economic or political crises, as has occurred in the former Soviet Union, whose once vigorous space program now rests on the hope of selling rocket launches to other countries. Even in times of relative stability, the public's frame of mind can prove fickle, the more so because space exploration offers economic rewards that remain questionable and in any case lie far in the future. The inconstancy of governmental and popular opinions with respect to exploring the universe reaches a maximum when it comes to sending humans into space—the most visible, most expensive, most dramatic, and for many people the only real part of the space program.

Consider the exploration of space by humans as it exists in the year 1999. The longest journeys through space have been eight different missions to or near the moon, with six lunar landings and two flybys. These missions, the culmination of the United States's *Apollo* program, all occurred between 1969 and 1972. The last person to walk on the moon left its surface more than twenty-six years ago. Astronauts have instead spent increasing amounts of time in orbit around Earth at altitudes no greater than a few hundred miles above its surface. John Glenn, the first United States astronaut to orbit Earth, became a national hero after his three circuits of the globe in 1962. Thirty-six years later, he waged a successful campaign to persuade NASA to allow him a final set of orbits on the Space Shuttle. This trip provided a spectacular finish to his distinguished career as a United States senator, but the fact remains that Glenn reached the same distance from Earth in 1998 that he did in 1962. This distance amounts to only one-thousandth of the distance to the moon, to one-millionth of the distance covered by the automated probes we have sent to Mars, and to one-hundred-billionth of the distance to the nearest stars.

As surely as we must learn to walk before we can run, short journeys must precede longer ones in our exploration of the universe. But we have in effect gone backward, from trips to the moon three decades ago to simple Earth orbits today. In hindsight, this strange sequence arose because we tried to run too soon. With the cold war at its height, the United States embarked on a race with the Soviet Union to send the first astronauts to the moon. Through its heroic expenditures, and because the Soviet Union abandoned its lunar astronaut program after it experienced repeated mishaps and incurred enormous costs, the United States won that competition. After the race was won, however, there was no justification for taking further steps in human space exploration. As the cold war approached its unforeseen end, Carl Sagan and other leaders in forming public opinion promoted the concept of a joint mission to Mars by the Soviet Union and the United States, but this notion fell into the dustbin of history, at least temporarily, when the Soviet Union collapsed at the start of the 1990s.

For now, human space exploration consists of Space Shuttle flights into orbits a few hundred miles high, along with a few Russian spaceflights into similar trajectories. The spacecraft involved have been designed to go no farther than these "low Earth orbits," which are important because they already take astronauts, and any equipment they may deploy, above nearly all of Earth's atmosphere. Many of their payloads have been used to study not the cosmos above but the ground below, or to speed communications around Earth. For these latter purposes, low Earth orbits are often superior to higher ones, but for further exploration, we must dream of longer voyages.

Could space tourism soon prove to be a paying proposition? Each Space Shuttle flight costs approximately $400 million and can carry as many as eight astronauts. Thus we can assign a cost to John Glenn's 1998 spaceflight somewhere between $50 million (if we regard him as one of the eight astronauts) and zero (if we take the view that adding one more passenger costs relatively little). If we reconfigured the Space Shuttle for tourism, it might take several dozen passengers into orbit. With ticket prices set at $10 million each, these journeys might show a profit as long as we can find passengers willing to pay these prices for a tour around our planet in rather cramped quarters. As a practical matter, space tourism must wait until we find ways to reduce the per-pound cost of spaceflight by a factor of at least a hundred.

Three Different Ways to Explore the Cosmos

To understand the big picture of space exploration, we must realize that we have three basic ways to extend our knowledge of the vast universe in which we live. First, we can build better instruments to study the universe and place them at superior sites for observation. The most notable and significant of these sites are locations outside our planet's protective veil of atmosphere, which blocks much of the information that would otherwise reach Earth's surface and hinders the remainder. Second, we can send automated probes to relatively nearby objects, which for forty years has meant objects within the solar system. Eventually, we may hope to send similar probes to other stars and their planets, which lie tens of thousands of times farther from us than the most distant worlds we have already visited. Third, human astronauts can go into space to examine its contents, bringing their eyes and minds to bear directly on new worlds, as long as we can provide these explorers with transportation and life support systems to ensure their survival. Being human, we tend to think of this as the height of exploration achievement, the continuation of an impulse that has spread humanity around the globe, to its mountain heights, to each of its poles, and to the depths of its oceans.

The first of these lines of attack, the quest to create better instruments and to place them at superior observation sites, has revolutionized our knowledge of the cosmos. Mammoth telescopes on high mountains, such as the Keck telescopes that peer at the universe from the summit of the Mauna Kea volcano in Hawaii, have joined spaceborne instruments such as the Hubble Space Telescope in securing sharper, deeper, and more precise views of the universe than any previous ones. Instruments in space have obtained images and made measurements that are simply impossible to perform even from high mountaintops. The Hubble Space Telescope, for example, studies the cosmos not only in familiar visible light but also in ultraviolet radiation, which cannot penetrate the atmosphere.

The second approach, carried out by automated probes throughout the solar system, has likewise yielded a flood of new results, though they refer directly only to our local set of objects and only by inference to the situation in other planetary systems throughout the Milky Way and beyond. In 1999 the *Galileo* spacecraft continues to sail on long, looping trajectories of observation around Jupiter, the largest planet in the solar system, obtaining detailed images of Jupiter and its four large satellites, especially of Europa, considered a likely site for life because it apparently maintains a liquid ocean beneath its icy surface. After August 1999, the *Cassini-Huygens* spacecraft, having received a "gravity boost" from its close passage by Earth during the summer, will spend four years on a voyage out to Saturn, nearly ten times as far as Earth is from the sun. There the spacecraft will detach a probe that will descend through the thick, murky atmosphere of Saturn's satellite Titan, where life may float in lakes of ethane and other hydrocarbons. Three spacecraft are now on their way to Mars to build on the success of the *Viking, Mars Pathfinder,* and *Mars Global Observer* missions; one of them will shoot a probe into the ground near Mars's south polar cap to measure the temperature, density, and soil composition.

In addition to these planetary missions, NASA's *Near-Earth Asteroid Rendezvous (NEAR)* spacecraft has made its first rendezvous with the asteroid Eros, a battered hunk of rock a few dozen miles across, left over from the formation of the solar system. In May 2000 this probe will match its orbit precisely to Eros's, with the spacecraft separated from the asteroid by only a few miles, so that *NEAR* can survey the pocked surface of Eros for clues to how the solar system formed. Another NASA probe, the *Stardust* spacecraft, left Earth in February 1999 on a series of orbits around the sun that will lead to a close encounter with the comet Wild 2 in January 2004. *Stardust* will not only make close-up observations of the comet but also acquire the first samples of cometary material, to be captured softly in an aerogel trap and returned to Earth in a re-entry capsule in the year 2006.

The spacecraft now at or on their way to Mars, Jupiter, Saturn, Eros, and Wild 2 have been designed and engineered to survive the hostile environment of interplanetary space, in which harmful radiation of different types creates a life-threatening danger to any unshielded astronaut. This fact emphasizes that the third leg of our triad of space investigation possibilities, human space travel, poses far greater problems—hence far greater costs—than the first two. Humans in space require much more than automated space probes do. Among these needs we should count not only air, water, food, and shielding against harmful radiation but also the much greater amounts of fuel required by a spacecraft with sufficient room and shielding for human activities, together with extra amounts of otherwise redundant strength and supplies to ensure a high probability of astronaut survival. Furthermore, if we expect the astronauts to return to Earth (and most of us, including most astronauts, will insist on this aspect of human spaceflight), all these requirements must be substantially increased to allow for a trip out and a trip back. Automated space probes make no such insistence; their loss, should it occur, may prove mysterious and highly frustrating to scientists who staked their careers on success but produces no human bereavement.

Who Will Choose the Best Ways to Explore?

In view of the imperious exigencies of human spaceflight, we cannot be overly surprised that astronauts have made only the first, tentative human voyages in space. A burning question remains before us: How should we distribute our efforts in cosmic exploration among the three possibilities—improved instruments for observation, space probes for close-up study within the solar system, and human investigation of celestial objects?

The brief answer to this question, historically speaking, is that we look to our elected representatives, and in particular to the United States Congress, to formulate policies regarding the funding of programs to improve our knowledge of the universe. Although private foundations continue to support important projects such as the twin Keck telescopes in Hawaii, the growing expense of these enormously capable instruments suggests that national governments should distribute the tax-enforced contributions from their citizens in accordance with the principles of a democracy. Indeed much of the funding for the second Keck telescope came from NASA, which hopes to use it in part to search for more planets orbiting other stars. For both automated space probes and human spaceflight, governmental support, either from individual nations or from nations that pool their resources for such expensive undertakings, has provided the organizational and financial backing necessary for these programs to succeed.

Must this be so in the future? In succeeding chapters, we shall meet scientists and engineers who disagree, almost to the point of anger, with the concept that governments should rule space. Instead, they argue that space belongs to the people and that the people, acting as individuals or through corporations, should explore, exploit, and indeed own celestial objects. For example, Arthur C. Clarke, the famous science fiction writer and visionary, supports the notion of private missions into space. The X-Prize Foundation of St. Louis, quite independent of any government, has offered $10 million to the first private group that twice successfully sends astronauts into space, three at a time, on separate flights no more than two weeks apart. Clarke sees this prize, like the $25 thousand prize that motivated Charles Lindbergh and his competitors to fly the Atlantic seven decades ago, as a useful incentive to help us to fulfill our nature as humans, which has always been, he says, "to explore our surroundings, to push the limits of our understanding, and to turn frontiers into future homes." Although Clarke neither explains why a prize would be needed if our nature demands that we explore, nor assesses whether sooner is always better than later in turning frontiers into homes, his attitude corresponds to that of large numbers of people who want a chance to ride through space and hope that private enterprise will provide what governments cannot or will not.

The clash between those who want governments to send astronauts into space and those who believe that private enterprise can do a significantly better job should raise a heartfelt reaction in many a reader. But the reader should address an issue still deeper than this one. To me, the argument over using governmental or private means to send human expeditions throughout the solar system amounts to debating whether to go to hell in a chemically- or nuclear-powered spaceship. We should ponder long and hard before we plant ourselves too firmly on other worlds. With my prejudices clear, I hope cheerfully to present opposing views in succeeding chapters of this book.

The International Space Station

For now, the farthest horizon where future homes for humanity are under construction lies a few hundred miles above us, at the largest building site ever opened in space. In November and December of 1998, Russia and the United States launched the first components of the International Space Station, an immense undertaking that will cost more than $100 billion, of which the United States will contribute a bit over 85 percent, with more than a dozen other countries supplying the remainder. This project, which we will simply call the Space Station, amounts to by far the largest construction project ever undertaken in space. In inflation-adjusted dollars, its estimated cost will fall a bit below that of the *Apollo* program that sent astronauts to the moon. The "bit" in question currently amounts to $15 billion, but the uncertainty of future costs could reduce this amount to zero or even to a negative number. By about the year 2008, the Space Station should be completed to the point that about a dozen astronauts can live in a low Earth orbit for as long as they like.

Because the Space Shuttle can carry astronauts no more than three hundred miles above Earth's surface, the Space Station will be built just below this altitude, about 285 miles high. Like all objects that move in orbit only a few hundred miles above Earth's surface, the components of the Space Station and the station itself take about ninety minutes to complete each circuit of our planet. As they do so, they are always falling not toward but *around* Earth. As a result of their continuous fall, astronauts in orbit feel "weightless," even though Earth's gravity continues to pull on them. A similar feeling occurs whenever you leap from a high diving tower: During your few moments of "free fall," Earth attracts you with its gravitational force, but you feel no effects from this force because nothing brakes your fall. Similarly, when everything in a spacecraft falls in the same direction and at the same speed, the effects of the Earth's gravity go unnoticed, even though the force of gravity continues.

Engineers involved in space projects sometimes use the misleading term "zero gravity" or "microgravity" to describe the situations in which a continuous fall produces weightlessness. But gravity acts on astronauts in near-Earth orbits as well as on those standing on Earth's surface. If a giant hand suddenly stopped the astronauts' motion in orbit and then let them go, they would fall straight down toward Earth's center. Conversely, the astronauts' momentum as they orbit would carry them straight off into deep space were it not for Earth's gravitational pull, which keeps them orbiting along a circular trajectory. The balance between gravity and momentum allows an object in orbit to continue orbiting our planet for centuries or millennia, neither escaping into deep space nor falling to Earth, until the extremely modest drag forces from the tenuous uppermost portions of the atmosphere eventually slow its motion. Eventually, any such object will enter the lower atmosphere, thus becoming subject to still stronger drag forces, and will finally strike Earth as an artificial meteorite, as NASA's *Skylab* space station did in 1979.

The International Space Station, which has taken more than a decade to plan and will require nearly as long for construction, has an orbit that should keep it safely above us for many centuries. To build and supply it, NASA plans more than a hundred Space Shuttle flights, eventually producing interconnected modules of laboratories and living quarters that will enclose a volume roughly twice the interior size of a Boeing 747 aircraft. (Following its own idiosyncratic accounting procedures, NASA estimates the Space Station's cost at $30 billion, rather than the actual $100 billion, by the simple technique of placing Space Shuttle flights essential to the project in a different budget.) This mammoth enterprise carries with it the possibility of accidents in space. The most significant disaster in past history was the explosion of the Space Shuttle *Challenger* on launch in 1986, which induced NASA to keep its other three Space Shuttles on the ground for almost three years. If such a tragedy should occur, the construction schedule for the Space Station will be similarly stretched out in time.

To build the Space Station, astronauts must overcome considerable challenges. Beyond the protective veil of Earth lies a cosmos hostile to their survival. Not only must they bring their supplies with them but they must also protect themselves against the host of particles and floods of radiation that continuously bombard our planet. Against the radiation, most of which comes from our sun (the more so during times of "solar storms"), astronauts can protect themselves with less than an inch of metal shielding—but must be sure to do so at all times, including during "space walks" (in NASA-speak, extravehicular activities, or EVAs) that are absolutely essential to assemble the Space Station. The particles that threaten a spacecraft or a space suit consist mainly of micrometeoroids, pieces of fine debris in orbit around the sun, left over from the processes that formed the planets, 4.5 billion years ago. Unless a particle's orbit precisely matches Earth's, its relative speed of impact on a spacecraft or astronaut will be large—as much as five miles per second, sufficient to transform even a tiny dust grain into a dangerous projectile. A micrometeoroid less than one-hundredth of an inch across can punch through the space suit of an astronaut on an EVA, and a dust grain just ten times larger can penetrate a steel bulkhead half an inch thick.

To protect astronauts against these cosmic bullets, which travel dozens of times more rapidly than the projectile from a rifle, NASA plans to reinforce the walls of the Space Station with layers of kevlar (best known for its use in bulletproof vests) and to develop quick-acting wall-patching kits to guard against a total loss of cabin air pressure. Ironically, the near-Earth environment, in which the Space Station will orbit, has become an especially dangerous locale because of the debris already left behind by previous space missions. A modest fleck of paint can damage a window on collision, while an item much larger than a micrometeoroid—a stray nut from an earlier satellite, for example—could make a sizable hole in even the strongest bulkhead the Space Station will have.

What long-term payoff will come from overcoming these dangers to build a successful Space Station? NASA once claimed that great benefits would flow to Earth's inhabitants, including new manufacturing processes carried out in the weightless conditions of space. The estimated time when these processes will prove cost-effective has receded far into the future, and NASA now stresses that the Space Station will provide increased experience in dealing with the problems and hazards confronting those who plan to spend long intervals of time in space. (On occasion, NASA's goals seem even more modest; thus in November 1998, when Russia launched the Space Station's first module, Dan Goldin, NASA's administrator, stressed its international aspect, which indeed would once have been unthinkable. Speaking in the Bronx accent that gives his utterances a street-wise knowingness, Goldin stated, "It's gonna be tough. It's not gonna be pretty. But we're going to have a *real* international space station." This remark epitomizes Goldin's ambivalence toward a project too far along to be stopped but which threatens to consume so much of future NASA budgets that more useful research may not go forward.) Although the Space Station is mainly producing debt for the public and profits for aerospace corporations, NASA was correct on one key point. The International Space Station will provide an excellent way—the best way, in fact—to discover how humans can survive for long periods in space. To the extent that this is our goal in space exploration, the Space Station can fulfill it.

NASA also points out that when we finally create a network of outposts in the solar system, a space station will be an immensely handy item to have. Even though the space station orbits only a few hundred miles above Earth, in order to join it in orbit we must expend about half the total energy required for a journey to the moon or to Mars. This is so because of Earth's significant gravitational force, which we effectively overcome by achieving an orbit that allows us to fall around Earth. From this position, we need only a relatively modest rocket to send a spacecraft onward, plus, of course, the ability to land on other celestial objects and eventually to return—not to Earth, but to the space station. For travel from the space station back to Earth, we can envision a fleet of specialized, short-hop rockets, like the switchyard engines of a long-distance railroad system. Eventually, of course, humans may travel back and forth from a near-Earth orbit to other worlds in the solar system without ever landing on our home planet—a plan that will save considerable amounts of energy but will first require the creation of a community in orbit around Earth.

However, not everyone interested in sending humans to explore the solar system favors construction of the Space Station. Many of the people who are most enthusiastic about sending astronauts back to the moon and onward to the planets find the emphasis and expense associated with the International Space Station rather ludicrous. To find out how to survive on a voyage to Mars, for example, we should create a mission to Mars, not a spaceship that orbits Earth seven thousand times a year and finds itself no farther away than when it began. Allowing for a certain willingness to risk the lives of the daredevils who might embark on the first trip to Mars, this argument makes reasonably good sense, and we shall consider it again in Chapter 5.

In assessing the difficulties of long-term human endurance in space, we must not pass lightly over the medical and psychological issues while noting the physical problems of food, water, air, fuel, and shielding. Four decades of sending humans into space have shown, unsurprisingly, that the sensation of weightlessness causes serious problems for organisms that have evolved in an environment with significant gravitational effects. Adaptation to weightlessness does not come easily. Intriguingly, some humans avoid the nausea of spaceflight far more easily than others, without any obvious advantages of youth or fitness. The weightless conditions in space move fluids from an astronaut's lower limbs into the head and also increase the amount of blood in the chest, fooling the heart until an innate regulatory mechanism decreases the plasma volume, simultaneously thinning the blood. Without gravity to fight against, muscles and bones quickly become weaker, especially in the body's major load-bearing structures. Many of these conditions can be simulated and have been experienced simply by lying horizontally in bed for several weeks without rising. With daily exercise, astronauts can avoid most of the changes to bone and muscle that would otherwise result from weightlessness.

Even more difficult than adapting to the weightless condition may be the readaptation to gravity conditions after landing, an important consideration for travel to other worlds. Astronauts returning to Earth have often found themselves disoriented for several days, like someone who arises from weeks of constant bed rest. We can anticipate, for instance, that astronauts who reach Mars will spend days or weeks recapturing their nonweightless selves before they can tolerate a planet whose surface gravitational force is about half that on Earth. The lunar environment, which the *Apollo* astronauts tasted only briefly, offers gravitational forces just one-sixth those on Earth—probably enough to keep the body happy on a long-term basis, and offering such new possibilities in athletic contests (once the problems of space suit bulkiness have been overcome) as the mile-long home run and the five hundred-yard football pass.

Other recreational activities will sooner or later lead to the first children conceived in space, and eventually to the first to be born in orbit or on a celestial object other than Earth. NASA's studies on mice and rats born in space imply that the absence of gravitational clues in infancy may have serious negative effects on adult animals. Much research remains to be done in this area before we celebrate the first space-born human astronaut. But once the first has been born, can the one millionth be far behind?

Are We Destined to Colonize the Cosmos?

From a long-term perspective, any argument over whether or not humans will live in space seems pointless. Judging by our past behavior, once we acquire the ability to establish ourselves on other celestial objects, some of us will do so. The issue then reduces itself to whether the human race will survive sufficiently long and will therefore prove itself able to develop the necessary technological skills for these colonies to spring into existence, to flourish, and to expand to the point where they can establish new colonies of their own.

In the final chapter of this book, we examine some potential flaws in this expectation and conclusion. For now, however, let us assume it is correct and that only the long-term survival of the human species calls into question whether we shall ever send colonists to other worlds. Were it not for the fact that this colonization cannot begin tomorrow, the sharp differences of opinion over whether such colonies are a good idea would rise to the forefront of public attention. As things are, however, we can leave these debates for our grandchildren, who will be the first to decide the merits of settling humans on the moon, Mars, asteroids, or the moons of Jupiter.

Or will they? A small group of technologists, dedicated to the goal of sending humans to live in space and convinced that this project can offer the inspiration and commitment that human society so desperately needs, have called for space colonization to begin immediately. To quote Robert Zubrin, one of the leaders in this effort, "We have in hand all the technologies required for undertaking within a decade an aggressive, continuing project of Mars exploration." Zubrin and his associates plan to accelerate their plans by avoiding governmental interference and direction; instead, they hope to obtain private contributions that will allow them to send humans to Mars, first to explore, then to colonize, and eventually to transform Mars into a planet more like Earth.

Listening to or recounting these plans, I tend to recoil in horror. I may not mind so much creating garbage on Earth, where we can hope to deal with it someday, but I draw the line at contaminating other planets. Among other problems, a tremendous contamination issue arises when dealing with a celestial object that might have life beneath its surface or might carry a fossil record of life in its rocks and dust. The thought that humans might expend enormous energy to reach the red planet, only to discover that Earth-born, Mars-borne life overlays and conceals any record that indigenous life may have left on Mars seems too sad to be endured. On a still deeper level, it is questionable whether we possess the moral right not simply to explore, which we can rather easily excuse as the inevitable product of our remarkable curiosity, but also to exploit, as, for example, by strip-mining the surface of the moon or of mineral-bearing asteroids. Depressing as this prospect appears, we descend yet one step further when we contemplate the possibility that faint-hearted, namby-pamby eco-nuts will crowd Earth while their bold and dauntless brethren carry the human flag toward worlds ripe for the taking, justifying their actions not only as entirely human but even as noble expressions of the private property concept that has served our society so well.

What forum exists in which to thrash out these competing views of space exploration? What process will or should lead humanity to a common view of the cosmos in which we live? What authority will enforce any decisions arising from our discussions of space and its contents? Quite clearly, the answer is none. In Chapter 4, we consider the role that the United Nations has played and could play in regulating human activities in space, but in this arena, as in terrestrial matters, we shall see that a great deal of effort would be needed to achieve even modest unanimity, which could be overthrown by a few enthusiasts with a mission. To be sure, we have no definitive proof that unity is desirable, and we all can imagine circumstances in which individuality should triumph over communal decisions.

So onward into space! Stay with me, and I'll take you to the moon and to Mars, outward through the asteroids to the giant planets and their satellites and then onward into the depths of the Milky Way galaxy in which we live. At each stage we will examine what seems feasible, and what may become possible in the future, for sending both automated spacecraft and human adventurers to distant celestial objects. I shall attempt to keep my prejudices subject to full disclosure, and to present opposing views of human destiny as fairly as I can. My goal is to bring us to the end of our mental voyage through the Milky Way wiser and more fully informed, ready to participate in the debates that will arise about what is to be done with the cosmos.

Chapter 2

The Great Leap Upward

The thousand-year history of rocket science on Earth began with the earliest known missiles, powder-packed tubes that were ignited and "launched" in China (apparently only for amusement) once the Chinese had learned how to make what we now call gunpowder. The western world eventually learned of gunpowder and quickly recognized its military importance, building cannons (first recorded in warfare during the first third of the fourteenth century) and steadily improving both their firepower and that of the projectile-firing rifled gun, the much-loved symbol of American freedom. A hundred and thirty-five years ago, as the American Civil War came to an end, Jules Verne wrote his novel *From the Earth to the Moon,* about an ex-Confederate gunnery expert who realizes his dream of building a cannon so large that it could send a projectile containing hardy human explorers on the quarter-million-mile journey to our satellite.

Leaving the Cradle

Verne made various scientific mistakes and glossed over other difficulties in the interest of moving his story forward. No technology of his time (or of ours) could protect astronauts from the mighty blast of a single explosion that could send a space capsule to the moon. The man who first saw how modern rockets could safely launch humans into space, Konstantin Tsiolkovsky, was a boy of eight when Verne's book appeared. Tsiolkovsky, who became a self-taught provincial schoolteacher in czarist Russia, conceived a desire (from infancy, he claimed) to avoid the bonds of gravity, which he saw as the oppressive enemy of humanity. In an essay written in 1883, he emphasized that the near vacuum of space imposes a requirement on any rocket to carry with it not only its fuel but also its oxygen, necessary to combine with the fuel in the combustion that propels the rocket. Tsiolkovsky also refuted the incorrect notion that rockets cannot function in space because a rocket needs something to "push against" and therefore will remain motionless once beyond the atmosphere. Instead, Tsiolkovsky pointed out, what counts is Newton's second law of motion: If you throw something in one direction, you tend to accelerate in the opposite direction. So long as a rocket can expel a stream of fast-moving gas, it can accelerate to progressively greater speeds.

These principles of rocket propulsion formed the centerpiece of Tsiolkovsky's masterwork, *The Exploration of Cosmic Space with Reaction Motors*, published in 1903, the year of the Wright brothers' first flight in an airplane. In a later work, Tsiolkovsky looked forward toward the day when space colonies would surround Earth and humans would establish colonies in the far reaches of the solar system. The Bolshevik revolution of 1917 brought a wave of modernizing enthusiasm well matched to Tsiolkovsky's visions, and in 1924 the Soviet government created a "central bureau" to study the problems and possibilities of rocketry, including military applications. Tsiolkovsky, who was pronounced a hero of the Soviet Union before his death in 1937, coined a two-sentence mantra that has reverberated among space enthusiasts throughout the twentieth century: "The Earth is the cradle of mankind. But one cannot live in the cradle forever."

Governments in the United States and western Europe showed far less interest in these possibilities. Appalled by the carnage of World War I, they left dreams of spaceflight to groups of private amateurs, who established the American Interplanetary Society in 1926, the German Verein fuer Raumschiffahrt (VfR) in 1927, and the British Interplanetary Society in 1933. Impressively, however, by far the most significant advances in rocketry during the period between the two world wars came not from any of these groups, let alone from a governmental institution. Instead, the man who turned space rocketry from theorists' insights into practical reality was an American born the year before Tsiolkovsky explained the principle of rockets, Robert Hutchings Goddard. Raised and educated in Worcester, Massachusetts, Goddard developed an early interest in spaceflight. After obtaining a Ph.D. from Clark University in 1911, he was hired there as an assistant professor and began his research into rocket motors, having already convinced himself that a rocket would work most efficiently if it carried liquid hydrogen fuel together with liquid oxygen to combine with the hydrogen in combustion. By 1918 Goddard had written a manuscript (which he ordered to be kept sealed until long after his death) predicting that great numbers of humans would live in space, drawing their raw materials from celestial objects in orbit around the sun, and would eventually escape the effects of the sun's death in what we now know to be about five billion years' time by riding nuclear-powered asteroids through the Milky Way.

Today we remember Goddard not for his dreams but for his practical accomplishments. In 1926, seven years after the Smithsonian Institution published his technical note entitled "A Method of Reaching Extreme Altitudes," Goddard gave a practical demonstration of his concept on his aunt's farm near Worcester, where he launched the first liquid-fueled rocket in human history. Like all of his later projects, Goddard's first flight involved a craft that could soar through the air without relying on atmospheric oxygen for combustion. Since the technology for putting small amounts of liquid hydrogen into a rocket did not exist, he used a more familiar fuel, liquid gasoline. Goddard's primary challenge was the development of a system to handle the liquid oxygen needed to make the gasoline burn, a liquid that must be kept at temperatures close to -300 degrees Fahrenheit to avoid evaporation. To pass this supercold liquid through a series of pipes and nozzles in carefully adjusted amounts and put just the right amount of oxygen into the combustion chamber, where it turned to gas and ignited with the liquid fuel, was a long technological struggle that Goddard successfully overcame.

During the late 1920s, eager to avoid what he imagined were prying eyes, and also to find better weather for a rocket testing ground, Goddard established himself in New Mexico. He was aided by small grants from the Guggenheim Foundation, which helped him despite the fact that *The New York Times* called him moon-mad, and by encouragement from the famous aviator Charles Lindbergh. Near Roswell, New Mexico, which later became famous as the alleged site of a UFO crash, Goddard built, tested, and launched a series of improved rockets, some of them twenty feet long and carrying several hundred pounds of fuel and oxygen many miles high. This work established him as the key pioneer in rocket development, an accomplishment that the United States government recognized in 1960, fifteen years after his death, when it paid a substantial sum to acquire his patents. A few years later, it named NASA's Goddard Space Flight Center in Greenbelt, Maryland, in honor of the man who launched the first liquid-fueled rockets.

Working in scientific independence of Tsiolkovsky's studies, Goddard recognized a crucial fact about how rockets could be used to leave Earth. Like Tsiolkovsky, he saw that what counts the most is the velocity that a rocket can attain and that the most efficient way to achieve high velocity is with a multistage rocket in which a smaller rocket sits atop one or more larger ones. As the lower rocket stages exhaust their fuel and fall away from the final stage, they give it extra speed, far more important than extra altitude. If the rocket's final stage reaches a velocity of about five miles per second, it can attain a semipermanent orbit around Earth; if it attains seven miles per second, it can escape Earth's gravitational effect and coast outward indefinitely. In contrast, a rocket that fails to reach five miles per second must fall back to Earth, and one with a velocity of less than seven miles per second, though it may have a large, looping trajectory, will remain permanently bound to our planet by Earth's gravity.

While Goddard labored in self-sought obscurity among his fellow Americans, the concepts he was exploring lit a fire in the hearts of enthusiastic Germans who banded together to build better rockets. The leader of this group,

Herman Oberth, born an Austrian citizen in Transylvania, was a science student in Munich when World War I began. Drafted into the Austrian army, he proposed that long-range, liquid-fueled rockets be constructed to rain destruction on the enemy. Although his notions were rejected as utterly absurd (just as well, or the descendants of Kaisers Wilhelm and Franz-Josef might still reign from their thrones in central Europe), Oberth's enthusiasm continued unabated. With the war over, he obtained a Ph.D. at the University of Heidelberg, where he encountered the ideas that Goddard had published in 1919. Like Goddard, Oberth resolved to devote his life to rocketry (not the usual sort of vow for a young man), and spent such funds as he could obtain on rocket research. Unlike Goddard, Oberth knew the value of promoting his work and vigorously sought to popularize it. He corresponded with Tsiolkovsky and others interested in rocketry, and in 1929 produced the first modern book on the subject, *Wege zum Raumschiffahrt* (*Paths to Space Travel*). When this book won a prize, Oberth promptly invested the money in further rocket research, and he became the established leader of the VfR, the Association for Space Travel.

From Amateurs to V-2s

Oberth's book suggested a new method of spaceship propulsion, which we now call an "ion drive." An ion drive strips atoms of one or more of their electrons, turning the atoms into electrically charged ions, and then uses an electrical field to accelerate the ions to high velocity. Oberth had no feasible way to make this dream a reality; instead, he and his VfR colleagues developed liquid-fueled rockets similar to Goddard's. Their efforts drew attention from the German army, which quietly supported the VfR even before Adolf Hitler came to power in 1933. Likewise, in the Soviet Union the government became interested in the rockets made by a band of private citizens who had formed a group analogous to the VfR, and in Great Britain a group of researchers founded the British Interplanetary Society to engage in rocket development but drew no more attention from the government than Goddard did in New Mexico. A group called the American Interplanetary Society was founded in 1930 to engage in rocket research but did relatively little in this regard and eventually transformed itself into a non-rocket-building association called the American Rocket Society.

Thus the 1930s saw unpublicized competition among three European groups and the American Goddard to produce bigger and better rockets, with the pride of a record high-altitude flight serving as the immediate prize. This record passed back and forth, from New Mexico to the Soviet Union, from there to Germany, and back to New Mexico in 1935 when Goddard launched a rocket more than 2.5 miles high. Soon afterward, with Hitler's rearmament of Germany well underway, the German army and air force decided to invest significantly in rocket research and set aside the island of Peenemuende on the Baltic Sea for rocket test flights. There an eager team led by Oberth devoted itself to designing, testing, and building instruments of military destruction that could bring the enemies of Nazi Germany to their knees.

Fortunately for the course of history, the Nazi rocket program was seriously curtailed after Hitler had a dream in 1939 about a rocket catastrophe, after which he directed that research moneys go elsewhere than the rocket program. Indeed, the Achilles heel of Nazi military research happily resided in the Third Reich's whimsical, mystical approach to decision making, in which a marvelous-if-true idea could usually supplant a reasonable, really good one because the former cost less and offered more. In his extraordinary book *Alsos*, Samuel Goudsmit, who led the key postwar investigation into Nazi science and technology, describes naval officers who used a swinging pendulum to locate enemy ships on a map of the Atlantic Ocean—far cheaper, if not as effective, as radar. The Nazis' nonscientific approach also lay behind their rejection of the atomic bomb as a concept worth sinking billions of marks into because far more destructive "death ray" types of weapons were suggested and not rejected as scientifically impossible. It also caused rocket research to limp along for more than two years, hiding its budget and barely able to build a single working example.

By late 1942, however, the dedicated Nazi scientists and engineers had successfully launched their first liquid-fueled military rocket, the A4. Unfortunately for them, however, grave quarrels broke out concerning the development of this weapon. According to a Gestapo report cited in Goudsmit's *Alsos*, "the Fuehrer called in all participants.... All military men were shown out within one or two minutes, because they were unable to answer the decisive questions of the Fuehrer. Dr. von Braun (technical chief at Peenemuende) was the only one who talked for thirty minutes and was able to answer the precise questions of the Fuehrer tersely and clearly. The Fuehrer decided, therefore, according to the proposals of Dr. von Braun." Soon after this conference, Hitler ordered full-scale production of the A4, which was given the sinister name *Vergeltungswaffe 2* (revenge weapon 2) or V-2 for short. From the concentration camps of conquered Europe, slave laborers by the tens of thousands were sent to V-2 assembly lines, which were moved into an underground facility in central Germany after a British air raid; there Hitler's victims died by the hundreds every day, making rockets that would rain down on England and destroy the "air pirates" (as Herman Goering called them) who had dared to drop bombs on Germany. During the year before Germany surrendered in April 1945, more than four thousand V-2 rockets were built and launched, causing several thousand deaths in Great Britain. This total fell far below that for the technologically simpler weapons called the V-1, which were essentially automated airplanes carrying large loads of explosives on preprogrammed (though rarely perfectly executed) flight plans.

The remarkably low casualty figure for the V-2 rockets—less than one death per launch—testifies to the rockets' poor guidance systems, which typically brought them down at distances of several miles or more from the prime target area. The design and construction of these guidance systems was, of course, in its infancy and remained a difficult task well after World War II had ended. The V-2 was a single-stage rocket, forty-six feet long, powered by liquid alcohol and carrying its own supply of liquid oxygen to be ignited with the alcohol in the combustion chamber. Carrying a payload of about one ton, the V-2 could reach a speed of about one mile per second, rise to an altitude of fifty miles, and reach a target two hundred miles away. The Nazi rocket team created a design for a multistage rocket with a range greater than three thousand miles, with which the dream of targeting the United States could be realized, and looked to even larger rockets capable of launching artificial satellites. These plans remained on the drawing board, thanks to the collapse of the German war machine in the spring of 1945.

The success of the German rocket program had left both the Soviet Union and its western allies keenly aware of the importance of capturing the V-2 rockets and, far more important, the men who had designed them. Aware of their value to their conquerors, the German rocket team, led by Werner von Braun, decided in January 1945 to surrender to the western forces. They buried their documents and, through what they no doubt considered heroic efforts, managed to reach the United States Army in Bavaria. The army in turn quickly plucked the documents from their hiding place before the Soviets could do so and then offered von Braun and his colleagues positions in the United States, where their expertise furnished an essential part of research into rocket flight for the next two decades.

Rocket Development in the United States and the Soviet Union

During the late 1940s and early 1950s, the United States used its stock of captured V-2 missiles as the basis for an experimental rocket program. This work culminated in the setting of a new altitude record with the first deployment of a two-stage rocket, as imagined by Goddard and Oberth decades before. In this effort, a small Viking rocket rode a V-2 to an altitude of fifty miles and then ignited its own motor as the V-2 fell away. The V-2 and Viking rockets made useful measurements in the upper atmosphere and obtained inspiring photographs, including the first images that clearly showed the curvature of

Earth. Their success demonstrated that launching a small Earth satellite would soon be feasible, and the Vanguard program was instituted with just that goal. Like the Viking, the Vanguard was designed purely for scientific purposes. By 1956, with its three-pound, softball-sized payload ready (but capable, with that weight and in that era, of making only the simplest of measurements and of broadcasting its presence and its results by radio), the Vanguard rocket neared completion for its launch from Cape Canaveral in Florida.

This scientific effort took place within a much larger, far more expensive and dangerous competition that had been underway for more than a decade since von Braun and his followers had surrendered to the western Allies. The success of the V-2 rocket had shown both the United States and the Soviet Union that ballistic rockets (the former name for what we now simply call rockets, and a reference to the fact that the rocket fired its motors for only a few minutes and then "went ballistic," coasting through the trajectory imposed by momentum and Earth's gravity) would play a determinative role in any future global conflict. The Soviet Union devoted a significant portion of its total effort toward manufacturing rockets capable of traveling the many thousand miles from launch sites in Russia to targets in North America, and reciprocal efforts were underway in the United States to design and construct missiles capable of striking the Soviet Union. This effort took more complex forms than its Soviet counterpart, in part because the United States had allies who could provide launching sites for relatively short-range rockets that could reach the Soviet Union, and submarines that could carry similar missiles close to Soviet shores. Another key aspect of rocket development in the

United States arose from the competition among the Army, Navy, and Air Force, each of which had separate programs to develop missiles, as well as a tendency to regard one another, rather than the Soviet Union, as the primary enemy. By the mid-1950s, each of the three major service arms had several missiles under development or actually under construction, including the Air Force's Atlas, with the longest range of all the different varieties, and the Army's Redstone, built by von Braun and his team at the Army Ballistic Missile Agency's main facility close to Huntsville, Alabama.

In the late summer of 1957, this intramural rivalry received a sudden jolt from an announcement by the Soviet Union that it had successfully fired a rocket over intercontinental distances, and a still ruder kick on October 4 when the world learned that not the United States but its chief competitor had launched the first artificial satellite of Earth. For most Americans, this news came as a terrible shock, a loss undreamt of in our worldwide struggle to be the best. The pain of being number two grew considerably deeper four weeks later when the Soviets put into orbit an eleven-hundred-pound satellite carrying a dog named Laika. (In those days, even in the United States relatively few people worried about the sad end that overtook Laika on her one-way space ride.) The president of Harvard University, along with other university presidents, called for an immediate increase in governmental funding to support higher education.

During the late fall and early winter of 1957–58, Vanguard rockets twice failed spectacularly to achieve success. In desperation, the government turned to von Braun's program, which in a brilliant, hurried effort produced a modest eleven-pound satellite, named it *Explorer,* and launched it with a Redstone rocket on February 1, 1958. Although *Explorer* restored some feelings of equilibrium, the "missile gap" between the United States and the Soviet Union suddenly became a key factor in the American political landscape as the realization sank in that our perceived enemies had much larger, more powerful rockets than we did, capable of sending a thousand-pound payload into orbit. While both countries strove to develop newer, bigger, and more accurate rockets, debate over the size and importance of the missile gap played an important role in the election of 1960 and continued to be a hotly contested issue for years thereafter.

In the spring of 1958, after *Explorer*'s modest success had underscored the United States' perceived inferiority in sending payloads into orbit, Congress created NASA. Because military applications of space technology seemed likely to play a determinative role in the struggle between the superpowers, and because military scientists and engineers had succeeded where civilians had not, it is not surprising that a great debate took place, mostly behind the scenes, as to whether NASA would proceed under military or civilian control. In the end, NASA was made an entirely nonmilitary operation, though no one could ignore the military implications of any technology that it might use or develop.

During the half-decade following the dawn of the space age in 1957, any efforts by the United States to put humans in space had to rely on the mighty Atlas rocket, the Air Force's intercontinental ballistic missile, which could rather easily be modified to carry an astronaut-bearing space capsule. This capsule and its rocket, named after Mercury, the messenger of the gods, rode the developmental fast track and seemed likely to restore the United States to the lead position in the space race—until April 1961, when the Soviet Union launched the first human to orbit Earth, Yuri Gagarin, and followed four months later with a second astronaut, Gherman Titov, who circled Earth for more than a day. The United States could come in only third, sending a chimpanzee named Enos on two Earth orbits in November 1961. Three months afterward, on February 20, 1962, John Glenn became the first American to orbit the Earth. His three orbits around the globe made him a hero and led him to a long political career as a United States senator who ended his term of service with a flourish by riding the Space Shuttle for about a hundred orbits of Earth in October 1998.

The Race to the Moon

In 1961 the *Mercury* program stood on the verge of success—but a success that still seemed to leave the United States number two in the space race with the Soviet Union. On May 25, seven weeks after Yuri Gagarin had become the first man in space, President John F. Kennedy, just four months after his inauguration, urged the United States Congress to fund a mammoth effort that would send American astronauts to the moon and bring them back to Earth by the end of the 1960s. Responding to the widespread feeling that the Soviet Union had adopted a similar goal, Congress enthusiastically agreed and the *Apollo* program was born.

Reaching the moon would require the development of a rocket with much greater thrust than the Atlas that had launched the first Americans—the mighty Saturn V, to this day the most powerful engine in its class. Before this rocket could be put into production, NASA used a series of modified Atlas rockets to launch *Gemini* spacecraft, each carrying two astronauts, into orbit. In 1966 the *Gemini* program achieved the first docking of two spacecraft in orbit—a maneuver essential to the eventual construction of any large structure in space.

The crucial components of the effort to achieve *Apollo*'s goal of a lunar landing were designed, fabricated, tested, and launched within a ring of southern states stretching from Texas, the home of the chairmen of both the House Space Committee and its Appropriations Committee and the newly chosen site of the Manned Spacecraft Center (now the Johnson Spacecraft Center), through Louisiana and Alabama, where the Army Ballistic Missile Agency became part of NASA's lunar effort, to Florida, whose good weather, low latitude, and proximity to a mostly empty ocean had already made it the obvious choice as the best site in the United States (except for the southern tip of the island of Hawaii) from which to launch rockets. The *Apollo* program thus produced a lasting effect on the economy of much of the United States' Deep South by stimulating the development of technologically oriented industries.

The goal of the *Apollo* program was to send a three-man capsule to Earth's natural satellite nearly a quarter-million miles away. There the *Apollo* spacecraft would fire its rockets to enter an orbit around the moon, from which it would send a detachable pod onto the surface carrying two of its astronauts while the third remained in orbit. The moon lander in turn would launch its upper portion back into moon orbit for a rendezvous with the orbiter, in which all three astronauts could return to Earth.

The *Apollo* program ranks among the most expensive undertakings of the United States government, costing well over $100 billion in current dollars. The government's decision to spend this enormous amount, while simultaneously expending huge sums on its military arm (with a major component of that spending devoted to the development of ballistic missiles), placed a strain on the economy of the United States. As the escalation of United States military efforts in Vietnam began in earnest during the second half of the 1960s, the political effects of these expenditures became a significant consideration. At the same time, the government of the Soviet Union decided that the expense of building a spacecraft to send humans to the moon was too great. Instead, the Soviet Union concentrated on automated spacecraft, as well as on the development of improved missiles. The governments of both the Soviet Union and the United States decided that neither could afford to lose superiority in building and deploying more powerful and more accurate rockets capable of destroying the other's society, an approach called mutual assured deterrence (MAD), which was much deplored but was still pursued by both.

The Quiet Success of Automated Spacecraft

Even with the *Apollo* program fully underway, the United States, like the Soviet Union, continued to develop improved automated satellites. By 1967 the United States had launched nearly five hundred automated satellites, and the Soviet Union nearly two hundred. Many of these spacecraft had a military purpose, most notably that of surveying enemy territory, but others have important civilian applications, most notably those of observing worldwide weather and relaying television broadcasts and telephone conversations. During the next decade, these numbers increased dramatically, especially as the result of efforts by the Soviet Union, which at one point in the 1970s was launching satellites at the rate of two per week.

Other countries in Europe, and Japan as well, developed their own automated satellites and launch capabilities, avoiding the mammoth spending needed to build rockets to send humans into space; their astronauts, however, joined those of the United States and the Soviet Union on many occasions. By the early 1980s, the major European countries had pooled their efforts to create a joint enterprise called the European Space Agency (ESA). Although ESA has a bureaucracy and a decision-making process that in some ways have proven even more cumbersome than NASA's (hardly unexpected when fourteen countries must reach an agreement on how to spend limited funds), both this agency and its Japanese counterpart, the Institute of Space and Aeronautical Science (ISAS), have designed and launched many successful space probes. Most notable among them have been the *Giotto* probe, which flew by Halley's comet in 1986, the *Infrared Space Observatory*, which investigated infrared radiation from a host of cosmic objects during the late 1990s, and the *Cassini-Huygens* spacecraft sent to Saturn as a joint undertaking of ESA and NASA.

Humans Reach the Moon

Despite the increasing number of automated spacecraft and despite the fact that the first space probes sent to other planets were already under construction as the 1960s opened, public attention understandably focused on what was perceived as a race to put humans on the moon. In this competition, the Soviet Union appeared to be winning all the preliminary phases. By 1959 *Luna I* had already become the first human-made object to escape the domination of Earth's gravity, passing within a few thousand miles of the moon on its way into interplanetary space. In May 1965, *Lunik 5* crashed onto the lunar surface after its soft-landing system failed, thus simultaneously becoming the first human object and the first human trash on the moon. Less than a year later, in April 1966, the Soviet Union sent a spacecraft into orbit around the moon, obtaining images of the lunar far side, the hemisphere that always faces away from Earth. Although the United States matched that feat later in 1966, in December of that year the world learned that the Soviet Union had achieved the first soft landing on the lunar surface with the spacecraft *Luna 13,* which examined soil samples and sent photographs back to Earth.

These successes not only seemed to rank the Soviet Union as the world's premier power in space exploration but also established it as the country most likely to send astronauts to the moon first. The United States' reversal of fortune in the competition for achievements in space came with completion of the *Apollo* program, which achieved nearly complete and stunning success during the late 1960s and early 1970s, proving to most of the public that United States scientists and engineers, fully supported by United States taxpayers, could aim for the moon and reach it. In December 1968, the *Apollo 8* mission sent three men around the moon and back again, our first emissaries to another celestial object. Then, on July 20, 1969, the *Apollo 11* mission touched down on the lunar Sea of Tranquillity, offering Neil Armstrong the chance to become the first human to reach another cosmic object in "one small step for a man, one giant leap for mankind." During the next few years, six more *Apollo* missions reached other parts of the moon; astronauts explored a few miles around some of the landing sites on an electrically powered vehicle and brought several hundred pounds of rocks back to Earth for careful study in our best laboratories. One of the spacecraft, *Apollo 13,* suffered a severe electrical accident soon after launch and survived only as the result of heroic efforts by the crew and those on the ground who oversaw the spacecraft's operations. (This episode, brought to the screen as *Apollo 13,* demonstrates that making Forrest Gump an astronaut is likely to produce serious problems.)

After winning the race to the moon, the United States' manned space program concentrated its efforts on long-duration flights in Earth's orbit. The *Skylab* space station, capable of supporting three astronauts for months, made its first appearance in orbit in 1973, the year after the last Apollo voyage. By the end of that year, astronauts had begun a *Skylab* mission that would last for eighty-four days, still the record for flight aboard a NASA spacecraft. Within a few years, the Soviet Union had put in place a still larger space station, called *Mir* (meaning both "peace" and "world" in Russian), on which astronauts of various nationalities have lived for periods exceeding one year. Although not specifically planned for this purpose,

Mir has provided an intriguing test of how close to failure a spacecraft can come without loss of life. Underfunded and woefully aged, the *Mir* spacecraft has undergone repeated power and computer failures, threatening disaster to the astronauts who happened to be occupying it. Quick thinking, hard work, and luck have averted catastrophe aboard the *Mir*, but these episodes remind us that space travel will remain a highly hazardous undertaking for years to come.

Thirty Years Later: Why Is It So Long Since Humans Have Walked on the Moon?

The *Apollo* program required a greater expenditure than any other operation in space to achieve its desired results, successful voyages by astronauts to the moon. The dark side of this project lies in its aftermath: Almost thirty years have passed since the last *Apollo* astronaut left the moon. Why?

The answer seems clear. The United States sent astronauts to the moon not as part of a carefully thought-out program of space exploration but in response to a perceived threat to the country's premier position in the world. The landing of humans on the moon was the first major step in space that the United States accomplished before its Soviet rivals. In fact, the Soviets had decided that sending astronauts to the moon was simply too expensive for them to undertake—an early example of a loss in economic competition that would culminate in the collapse of the Soviet Union during the early 1990s. But once the (actually nonexistent) race was won, there was no reason to go farther into space—no reason, that is, sufficiently compelling to galvanize Congress into action. NASA did develop other spacecraft to orbit Earth, including link-ups with Soviet astronauts in the *Apollo-Soyuz* program and long-duration flights aboard *Skylab,* the heaviest payload ever launched. However, in the argument between those who advocated spending hundreds of billions of (1999) dollars more to send astronauts to Mars and those who believed that human spaceflight was simply too expensive an enterprise on which to spend more than a few tens of billions of dollars, the conservative spenders carried the day, limiting all human activity in space to Earth-orbiting missions during the 1980s and 1990s.

By comparison, automated spaceflight has grown progressively. We have sent spacecraft to all the sun's planets except Pluto; some of these spacecraft have landed on planets, investigating the immensely hot surface of Venus, hidden from view by its opaque atmosphere, and the frozen plains of Mars, where primitive forms of life may have found ways to persist in crevices, at the polar caps, or in underground oases. We have photographed Mercury, Jupiter, Saturn, Uranus, and Neptune, as well as the satellites of the latter four planets. Even now, the highly successful *Galileo* spacecraft is completing its fourth year of operation around Jupiter, where it has repeatedly obtained images and made measurements not only of the sun's largest planet but also of its four great moons, Io, Europa, Ganymede, and Callisto, each as large as or larger than our own satellite. On its mission to Saturn and its large moon Titan, whose surface is judged to be the most favorable in the outer solar system for the existence of life, the *Cassini-Huygens* spacecraft will pass by Earth in August 1999, gaining sufficient energy from the "gravitational slingshot effect" to achieve a trajectory that will take it to Saturn, nearly a billion miles from Earth, in the year 2004.

All this exploration has occurred while astronauts have ridden the Space Shuttle or its Russian counterpart several hundred miles up, several dozen times around the Earth, and several hundred miles down again. We can regard this contrast between automated spaceflight and human spaceflight in at least two ways. One point of view concludes that humans cost a bundle, can survive in space for only relatively brief periods, continually risk their necks, and represent a burden on the prospects of funding true space exploration, which will be accomplished by robotic spacecraft. The other point of view regards automated spacecraft as the modest forerunners and advance guard of eventual human explorers, doing the reconnaissance work that will allow astronauts to make their voyages more efficiently and successfully as they spread human intelligence throughout the solar system.

To some extent, both views are correct; more precisely, each has its merits. Automated spacecraft will always precede human explorers, and although robots can perform a host of tasks quite well, there will always be others that humans can do better. The cost of human travel will always far exceed that of robotic exploration because any cost savings on launch and transport will apply to both and humans will always demand much more in the way of safety and provisions. As we shall see in the next five chapters, we also face a conflict as to whether government or private enterprise furnishes the best means of advancing space exploration. From the tension generated by discussions of automated versus human spaceflight and private versus public funding, we can hope to extract the best approach to the problem that humans can produce. We begin our careful consideration of these issues with a look at Earth's closest neighbor, the celestial object that we call the moon.

Chapter 3

Colonizing the Moon

Of all the predictions that might be made about future human exploration of our cosmic environment, one of the safest is that we shall someday establish a colony on the moon, the closest significant celestial object and the only object beyond Earth on which humans have already landed. The red planet Mars calls to us far more dramatically, with its tales of possible life hidden below its surface, than the moon does with its barren, rocky surface devoid of all but the most transient atmosphere. But even when Earth and Mars reach their minimum separation as they orbit the sun, Mars lies 125 times farther away from us than the moon's distance of 240,000 miles, which varies by only plus or minus 5.5 percent as the moon orbits Earth.

Fundamental Moon Facts

A diameter of 2,150 miles gives our satellite a surface area equal to about 7 percent of the Earth's. However, because water covers nearly three-quarters of our planet's surface, the moon's area amounts to more than a quarter of the total land area on Earth. The lunar surface displays two distinctly different types of features, ancient lava plains and still older highlands. The lava plains, also called seas or maria (the Latin word for "seas"), apparently arose during the great era of bombardment that dominated the solar system soon after it formed, about 4.6 billion years ago. The first half-billion years after the formation process brought a rain of objects impacting on the moon. Some impacts spread molten rock for hundreds of miles, burying the ground beneath these lava flows while the highlands remained basically unchanged. Four billion years ago, this destructive cosmic rain gave way to the gentle sprinkle of meteoroids that now continuously strike Earth and the moon, with a rare, much larger impact added for excitement.

Frozen lava plains now cover about one-third of the moon's near side, while highlands form the remainder. In contrast, the lunar far side has no lava plains, only highlands. To explain this difference, we should define the terms "near side" and "far side." The moon takes just under one month (called a "moonth" in Old English) to orbit once around Earth. (More precisely, the moon orbits around the center of mass of the Earth-moon system, but since Earth has 81 times the moon's mass, the center of mass lies inside Earth, though not at its center.) As it moves in orbit, the moon always keeps the same side facing Earth. We call this face the moon's near side and recognize some of its larger features as the "man in the moon." For untold millennia, only the moon's near side showed itself to human eyes; then the advent of the space age gave automated cameras and then astronauts the opportunity to gaze on the far side. The absence of maria on the lunar far side strongly suggests that the moon has apparently kept one side turned toward Earth since a time soon after its formation. Once that happened, Earth's gravitational effects on large objects striking the moon apparently diverted most of the largest of them onto the moon's near side.

And why does the moon always keep one side facing Earth? Subtle but significant differences must exist in the density and composition of the lunar interior, with the result that our planet's force of gravity has locked onto our satellite's denser regions, twirling it in a gravitational waltz as Earth and the moon orbit the sun. Other satellites in the solar system, such as Jupiter's four large moons, exhibit the same behavior: Each of these objects rotates in the same amount of time that it takes for a complete orbit, and so it always shows the same side to its planet. For our moon, this matching produces a rotational period of 27⅓ days, much longer than Earth's daily rotation. As the moon rotates, it presents its near side and far side to the sun in succession. The changing amounts of solar illumination of the near side produce what we call the phases of the moon, varying from full moon, when sunlight shines on all of the near side, to new moon, when sunlight strikes only the lunar far side. Every location on the moon except those close to the two poles experiences two weeks of extreme cold and darkness followed by two weeks of intense heat and light. During a few hours of transition from night to day, the moon's surface temperature increases from -180 degrees to 240 degrees Fahrenheit, an extremely large range that will cause a problem for future lunar colonists.

Despite its enormous fluctuations in temperature, the moon offers some impressive advantages for its occupants. First, the fact that the lunar far side never faces Earth offers a chance to avoid some of the scientifically negative effects of human civilization. The flood of light, radio waves, microwaves, and other forms of electromagnetic radiation produced by terrestrial needs for illumination, radar monitoring, and communications threaten to drown astronomers in a sea of local noise, vitiating their attempts to observe the universe and its faint (because they are so distant) natural sources of radiation. For example, if we hope to detect other civilizations in the Milky Way by "eavesdropping" on radio and television signals that they may leak into space (see Chapter 11), we must find a means of overcoming our tendency to discover ourselves because our leaked radio signals create a background of noise much stronger than any signal we might reasonably hope to detect. The calmest, quietest places in our local environment, protected from the light and radio pollution produced by humans, lie on the far side of the moon, always facing directly away from our home planet. Transportation to this locale, and communication with it, will inevitably cost more than travel to, or talking with, the Earth-facing side—a fact that restricted all six *Apollo* landings to the near side. Nevertheless, the opposite side of the moon offers such impressive advantages that the center of the lunar far side should become the prime location for a scientifically oriented lunar colony, where astronomers of the twenty-first or twenty-second century will make precise observations of the cosmos almost free of interference from Earth.

Problems Confronting Those Who Build a Lunar Colony

What will it take to create a lunar outpost? As we have seen, the basic human requirements for a space colony include oxygen to breathe, water to drink (and possibly to bathe in), food to eat, shelter against an environment far different from that which humans have evolved to enjoy, and sufficient motivation to survive hostile conditions indefinitely. The creation of a base on the moon, or anywhere else beyond Earth, resembles an attempt to live for years on the summit of Mount Everest or miles below the ocean surface—but many times more difficult.

Some of the hostile features of the lunar environment amount to a straightforward multiplication of the hostile conditions on a high mountain peak. Like climbers who venture onto the slopes of Everest, with its nighttime temperatures of -70 degrees Fahrenheit, moon dwellers must find a way to stay warm at -180 degrees Fahrenheit. In addition, lunar colonists must just as surely find ways to stay cool during the 240-degree-Fahrenheit day. Like high-altitude climbers, moon colonizers must initially bring all their food with them, along with oxygen to breathe. In addition, the lunar setting, like the conditions encountered in near-Earth orbit, embraces another set of dangers, one only partly experienced on the slopes of high mountains: exposure to deadly radiation.

Here the word "radiation" describes two quite separate hazards. One is the sun's short-wavelength ultraviolet radiation, capable of damaging fragile tissue, especially human skin and eyes. Our atmosphere blocks most solar ultraviolet radiation, but at higher altitudes, as the air grows thin, this protection decreases. At Everest's altitude, the atmosphere blocks ultraviolet radiation only one-quarter as well as it does at sea level, and so those who climb the mountain experience an ultraviolet danger closer to that confronting lunar prospectors than to that encountered by the rest of the human population.

Light waves, ultraviolet, and other forms of electromagnetic radiation consist of streams of photons, particles with zero mass, which nonetheless carry energy—energy recorded by our eyes when we see the world, or by our skin when we tan it, burn it, or simply degrade it under the impact of ultraviolet radiation. In contrast, the word "radiation" also describes other, potentially deadly, particles that permeate interplanetary space. These other particles have mass, unlike the photons that form electromagnetic energy. They are mainly protons and electrons, most of which come from the sun, which emits high-velocity streams of these elementary particles in a continuous outflow called the "solar wind."

Because protons and electrons have mass, they are more penetrating and carry larger amounts of energy than the massless photons that constitute ultraviolet radiation. No single particle offers much danger, but they strike by the millions, all the time, and their cumulative effects can be deadly. Earth's atmosphere blocks essentially all of these particles, not just at mountaineering altitudes but dozens of miles above its surface; even climbers who reach record heights have nothing to fear from them. Beyond the atmosphere, however, we enter a different regime, one in which everyone who ventures outside a protective habitat must work and play inside a spacesuit capable of absorbing the impacts of relatively high-energy particles. From time to time, "solar storms" produce especially large numbers of solar wind particles with particularly high energies. At such times astronauts should remain indoors, behind thicker shielding than a space suit can offer to block the impacts of the fast-moving particles from the sun. Because solar storms can be observed on the sun several days before their particles reach Earth and the moon, lunar colonists should have ample warning that they must enter one of these down time periods.

The enormous task of establishing a lunar colony will start with repeated landings on the lunar surface, most of them cargo flights to carry the material essential for survival: stores of food, tanks of water and compressed oxygen, power cells, and equipment for working on the moon. One of the first tasks of the advance teams will be to link empty fuel tanks of the cargo rockets together with prefabricated modules, creating a livable habitat from which colonists can venture forth to work on the lunar surface. With this habitat in place, the longer-range task of the colony will be the excavation and construction of an underground living space, offering protection against all types of radiation. In addition, the thermal insulation provided by the lunar soil will significantly reduce the need to heat the colonists' living quarters during the lunar night and to cool them during the lunar day.

In construction terms, the moon's resources stand head and shoulders above that of interplanetary space, a near emptiness devoid of raw materials from which to build a home. Basic home construction material lies all over the moon: rock and lava (once-molten rock). To build a suitably shielded lunar habitat, colonists need only dig into the moon, either to live below its surface or to create stone houses. The latter effort will not yield blocks of limestone or marble; much of the lunar surface consists of "friable" rock, which crumbles easily and cannot bear much pressure. Initially at least, lunar colonists are likely to dwell in artificial caves, emerging onto the lunar surface for carefully monitored periods of activity.

The moon will offer one significant advantage to these colonists, already enjoyed by the *Apollo* astronauts. On the lunar surface, gravity pulls with only one-sixth the force that we feel on the surface of Earth. Since we use the word "weight" to describe the strength of the gravitational force on an object, we can correctly say that each of us will weigh only one-sixth as much on the moon as we do on Earth. Hence even within the rather cumbersome space suits we would need to protect us from harmful radiation, we could skip and bounce over the moon's surface, engaging in various athletic feats that would be simply impossible on Earth. Lunar gravity should be sufficient to avoid the negative effects of weightlessness, which affect astronauts who spend weeks or months orbiting Earth. In time, the moon might prove a useful sanitarium for those who feel Earth's gravity all too keenly. Before that can occur, however, we must find ways to lower the cost of transportation to and from the moon.

New Techniques for Sending Material to the Moon

The early epochs of lunar colonization will require the influx of vast amounts of supplies from home, ranging from the bare necessities of food, water, and oxygen to the sophisticated equipment that may eventually render lunar colonies independent of Earth. To maintain a steady spacelift for these necessities, we must develop a transportation system far more efficient than our chemically fueled rockets can provide. With our present launch systems, each pound sent into orbit around Earth costs about $10 thousand, and each pound sent to the moon (not that we have sent much recently) costs a few times as much. To send a lunar colony's bare minimum of a million tons of supplies would therefore cost more than $10 trillion. Clearly no such expenditure will occur; colonies on the moon must wait for the time when we can reduce the per-pound cost of transport by a factor of one hundred to one thousand.

This may not take long, at least for nonhuman cargo. Demonstration models of systems that are capable of rapid development into much larger production models already exist, so long as we are launching only inanimate objects, which can withstand much greater acceleration than humans or other animals can. Though chemical rockets develop enormous amounts of thrust, they apply this thrust over a period of several minutes, producing accelerations that rarely surpass ten times the benchmark acceleration, called one G, that describes how rapidly an object near Earth's surface accelerates as it falls downward. Humans can withstand accelerations of five Gs or so, though they are not happy while they do. In contrast, the inexpensive launch systems that might supply a lunar colony would produce accelerations of hundreds or even thousands of Gs. These systems would fling packages directly from Earth's surface into a trajectory carrying them toward the moon, where other systems, still in the design phase, would brake their fall. Two good methods for achieving tremendous accelerations have been developed in prototype. One uses electromagnetic forces to accelerate its payloads, with a series of superconducting magnets, each of which adds its acceleration as the payload passes. The other system is a ramjet, a pipe that burns combustible gases behind the payload in a carefully designed manner so that the gas itself does not move but the payload accelerates to progressively higher speeds, emerging from the end of the pipe as a projectile on its way to the moon.

Any new space launch system will require an enormous investment in new designs and newly engineered technology. If the technology becomes a standard mode of operation, we may hope for—but may not necessarily achieve—striking savings in its employment. This was the path envisioned for the Space Shuttle, the first reusable vehicle to carry people and payloads into space. But the Space Shuttle has yet to demonstrate that it saves money over the older approach of launching a spacecraft only once. To begin with, the Space Shuttle is not entirely reusable: The two mammoth fuel tanks that give each departing shuttle much of its thrust on launch fall away after a few minutes and have often been lost in the Atlantic Ocean. More significant in adding cost, each launch and return of the Space Shuttle must be followed by extensive checking and repair of the vehicle. The bottom line is that each Space Shuttle flight continues to cost many hundred million dollars, about the same price as a "disposable" launch vehicle. The Space Shuttle may eventually pay off, and it seems destined to be the chief means by which humans will be sent into space for the next decade at least, but we have not saved much money, if any, in spending tens of billions of dollars on its development and deployment.

Magnetic or ramjet launchers for inanimate payloads should cost far less than the Space Shuttle, in large part because we risk relatively little if one or more payloads should be lost. Furthermore, any lunar colony will require many thousands of payloads to be sent from

Earth, offering significant savings once we have a reliable launch system in place. This number of supply launches will far exceed the hundred or so Space Shuttle flights that will be needed to build the International Space Station, or a similar number of rocket flights that could put astronauts on the moon with their initial supplies.

Empowering Lunar Colonies

Before a lunar colony could become largely self-sustaining, it might require that its food, oxygen, and water be furnished from Earth for decades. For the colony's power supply, however, we can reasonably hope that a few years would suffice to create self-sufficiency among moon dwellers. Two serviceable sources of energy head the list: nuclear power and solar power.

Nuclear power, familiar to us on Earth for the controversies it provokes, relies on the heat produced by radioactive minerals, usually isotopes of the elements uranium and plutonium. Each isotope of an element consists of slightly different nuclei, the combinations of protons and neutrons that form the heart of all atoms. With time, the radioactive isotopes "decay," changing into other types of nuclei and releasing heat as they do so. This heat can be captured and used in a variety of ways, typically in making steam that runs a turbine to produce electrical power. The term "nuclear power" refers to the fact that the heat comes from the decay of radioactive nuclei; another use of this phrase appears, as we shall soon see, when we encounter nuclear fusion, the melding of two nuclei.

The decay of uranium or plutonium isotopes can provide significant amounts of power from relatively small masses of radioactive material. Even though we must invest substantially in creating a mechanism to control and to use the heat released by radioactive decay, we may profit thereafter from the fact that we must resupply the power plant with only modest amounts of fuel, either sent from Earth or, in the more distant future, obtained from mining operations on the moon. Nuclear power poses the problem of dealing with radioactivity in the spent fuel, which emits dangerous radiation long after its commercial value has fallen to zero. We know from our own lives that no one wants this matter buried in his backyard. Hence we can count ourselves fortunate in leaving this problem to future lunar colonists, who will presumably accept the risks of moon life knowingly and may well judge this one to be small in proportion to the others. These colonists might in time develop rockets to send spent radioactive fuel into the sun, the safest way to dispose of it. (Safe, that is, as long as we can avoid any accidents in launching the plutonium—a major reason why we do not employ this method on Earth.)

Solar power offers a completely safe means of furnishing a lunar colony with heat, light, and ongoing machine operation. The moon, always at nearly the same distance from the sun as Earth, receives substantial amounts of solar power every second on its sunlit side, but none on its dark side. As the moon rotates, these sides slowly exchange positions on the lunar surface. Thus although for two weeks a lunar colony can bask in solar power that arrives unimpeded by clouds or other atmospheric effects, the following two weeks pose a serious problem, with not a photon of sunlight directly available. A lunar colony that runs on solar power would require a well-engineered system of batteries or flywheels that can store power during each two-week period of abundant sunlight and release it during the next two weeks as the cycle continues.

Advantages of the Lunar Poles

Two places on the moon do not obey the rule of alternating feast and famine in available solar power. These are the north and south poles of the moon, which may prove by far the most favorable lunar real estate for sizable numbers of inhabitants. Because the moon's axis of rotation is nearly perpendicular to the plane of the moon's orbit around the sun, a person at one of the moon's poles would see the sun rise only slightly above the horizon, take two weeks to reach its maximum elevation of less than two degrees, and then spend two weeks dipping just below the horizon. As a result, the surface temperature at the moon's poles varies through a far lesser range—no more than plus or minus 50 or 60 degrees from its average of about -100 degrees Fahrenheit—than the temperature does on the remainder of the moon's surface. More important than avoidance of the extremely cold nighttime temperatures that occur on the rest of the lunar surface is the fact that we could easily build solar collectors, rising sufficiently high that they remain in sunlight throughout the cycle of night and day. These collectors would avoid the energy storage requirements of a solar-powered colony elsewhere on the moon.

This advantage, though significant, blanches in comparison with a recently discovered polar advantage undreamt of until 1998, one to be grasped with vigor if it is true. In 1998 the *Lunar Prospector* spacecraft, in orbit sixty miles above the moon's surface, used its "neutron spectrometer" to detect the neutrons that are released when extremely fast-moving particles bombard matter. We can compare a neutron spectrometer's function to the study of matter by shooting bullets of a known type at material of unknown composition. The type of material under bombardment determines both the speed and the composition of the emerging fragments. On the moon, the "bullets" arrive continuously in the form of electrons, protons, and other particles in the solar wind, and of similar fast-moving particles produced in the depths of space. The neutrons detected by the *Lunar Prospector* are the fragments, which reveal the compounds that form the lunar surface and subsurface.

Lunar Prospector discovered that ice lies beneath the craters close to the moon's poles. The proportion of this ice varies from 0.3 to 1 percent, and so we must imagine not sheets of ice but ice mixed together with much greater amounts of rock and soil. This ice has remained in the soil, rather than evaporating into space, because the sun never shines directly onto the floors of craters in the polar regions. These sunshielded crater floors are the coldest spots on the moon's surface, kept at temperatures of nearly -400 degrees Fahrenheit. The ice on the moon, frozen within the soil below these craters, represents a relatively tiny amount of the moon's original water, located in the only regions where it could remain frozen for billions of years. From a space colonization viewpoint, nothing is better than ice, except for liquid water itself. We can melt ice to drink and bathe in liquid water, and we can separate water into hydrogen and oxygen and then use the oxygen both to breathe and also to combine with any fuel that we can extract from the lunar surface.

Though the discovery of ice at the lunar poles has energized those who plan for future colonies, we must not go overboard. Before we send missions to the moon's poles to extract the ice that nature has left there, we would do well to run the numbers. Current estimates of the amount of ice buried in the soil and revealed to *Lunar Prospector* reach tens of millions, perhaps several hundred million, tons. If we use this ice solely for drinking water and for oxygen to breathe, that amount could satisfy human requirements for perhaps one hundred million person-years. This calculation relies on the fact that a ton of water, most of whose mass consists of oxygen atoms, can supply an individual with a few months' worth of drinking water and breathing oxygen. Thus ten thousand lunar colonists could last for ten thousand years before exhausting the lunar ice, assuming that it could all be extracted and converted into water and oxygen. Lunar colonists would of course consume this water at a greater rate if they use it for agriculture in hothouses, and still more rapidly if they use oxygen to burn fuel.

Ten thousand people for ten thousand years, or one million for a century! These numbers may seem impressive, but they amount to a drop in the bucket when we consider the ultimate goals of lunar colonization. We ought to think twice before planning to mine the lunar poles to extract the ice in the soil, leaving us with nothing to drink when the water runs out. We must also confront the fact that this ice lies dispersed over roughly ten thousand square miles of lunar surface at each pole. To use it, we must therefore design and build machines to till these expanses of lunar soil and to refine the ice from a quantity of rock at least a hundred times greater. In view of the limited supply of lunar ice (at least when colonists number in the many hundreds or thousands), we might do well to set aside the moon's poles as an inner solar system ice reserve. Even this could be seen as arrogant by other civilizations who might have an eye on the moon's ice, but fortunately (from this angle at least), we know of no such other civilizations.

Growing the Lunar Population

In the medium term, which could mean by the end of the twenty-first century, a lunar colony might well attempt to become largely self-sustaining, using solar power for heating, cooling, and the separation of lunar soil into the materials needed for life, such as hydrogen, oxygen, and various metals. Is this only a dream? Or, conversely, have we made an overly conservative estimate of how rapidly humans can become self-sustaining on the moon?

It seems reasonable to anticipate that a hundred years from now, humans will be living on the moon, mainly to perform scientific tasks but also to develop the means of long-term survival on Earth's natural satellite. As we have noted, the lunar far side offers the best scientific site for study of the cosmos because the moon's bulk stands between that site and our planet, protecting it from Earth's immense output of noise. The gravitational force at the moon's surface, one-sixth as great as that on Earth, does not offer much of an impediment to construction projects, and the fact that we do have a surface on which to build makes many conceivable observatories far easier to fabricate and to maintain than would be possible in space. We can therefore anticipate human activity on the lunar far side and at the lunar poles, linked by a communications net, perhaps of moon-orbiting satellites that also relay data to and from Earth.

Helium 3, a Possible Lunar Treasure

Before we leave the moon to embark on more general considerations of space colonization, let us note that the lunar surface contains an extremely valuable resource, a rare variant of familiar helium gas: the isotope called helium 3. Ordinary helium, helium 4, consists of nuclei each made of two protons and two neutrons, with two electrons in orbit around the nucleus. Helium 3, in contrast, has two protons and *one* neutron in each nucleus, along with the two electrons in orbit. Even though both isotopes of helium appear only in the gaseous state (because helium atoms, which form an inert gas, refuse to combine with any other atoms under all but the most extraordinary conditions), small amounts of helium gas are almost certainly trapped inside lunar rocks. This helium consists of both helium 3 and the much more abundant type of helium, helium 4. Both types of helium nuclei reach the moon in the solar wind, which, as we have seen, does not penetrate to Earth's surface. In the solar wind, about one helium nucleus in every ten thousand is helium 3.

And what does this helium 3 offer us, assuming that we can extract helium gas from the moon's rocks and separate the helium 3 from its much more abundant cousin, helium 4? Unlike nuclei of helium 4, helium-3 nuclei readily fuse together. These nuclei do repel one another, as do any two nuclei, because they each carry a positive electric charge and like signs of electric charge repel each other. But if the nuclei are shot at each other so rapidly that they approach within an extremely small distance—about equal to the size of one of these nuclei, less than one-trillionth of an inch—then another force, called the strong or nuclear force, will cause them to cling together to form a new, larger nucleus. Two nuclei of helium 3 that fuse together typically form a nucleus of helium 4, plus two protons. This fusion also releases significant amounts of kinetic energy, typically in the form of photons with extremely high energies, called gamma rays, and fast-moving electrons and antielectrons. We can also fuse helium 3 nuclei with the isotope of hydrogen called deuterium, or hydrogen 2, in which each nucleus consists of a proton plus a neutron, rather than the bare proton that characterizes ordinary hydrogen nuclei.

The fusion of helium-3 nuclei, either with other helium-3 nuclei or with deuterium nuclei, closely tracks what happens at the centers of stars, where nuclear fusion produces the energy that allows the stars to shine for billions of years. Similar fusion reactions occur in the detonation of a "hydrogen bomb." Thus our own star, the source of light and heat on Earth, as well as the most powerful explosions that humans have created, involve nuclear fusion, including the fusion of helium-3 nuclei to produce other nuclear varieties.

Unlike most other examples of nuclear fusion, the fusion of helium-3 nuclei does not release fast-moving neutrons, which create a source of danger to those in proximity to the site of nuclear fusion. Thus the nuclei of helium 3 that lie trapped in rocks on the moon offer us a most useful, compact "fuel"—provided that we can arrange for the energy to be released in nuclear fusion not all at once (as occurs in a bomb) but in a controlled fusion reactor. For more than three decades, scientists and engineers have attempted to successfully design and construct such a reactor; they often seem on the verge of achieving success but cannot quite do so. The fusion inevitably produces gas at such high temperatures that no conventional container, no matter how strong, can escape melting. The hot gas must therefore be confined by magnetic fields, and although this is theoretically possible, the gas tends to slip out almost instantaneously between the lines of magnetic confinement.

When we do develop controlled fusion reactors on Earth, one source of fuel will be nuclei of hydrogen 2, called deuterium, which can be found in small quantities in sea water. An even better fuel supply, however, might be helium-3 nuclei extracted from lunar rocks. This would certainly be true if we plan to build a reactor on the moon, and because helium 3 fuses so efficiently and without emitting dangerous neutrons, it might even prove economically worthwhile to bring helium 3 from the moon to Earth. However, for every ounce of helium 3 to be extracted on the moon, we must dig up and process millions of tons of lunar soil. Will this really be a good idea? Within a century or so, human society will confront and presumably will answer this type of question, as lunar colonies require ongoing supplies of energy.

Will the Moon House the First Space Colonists?

As we leave the moon and its possibilities for colonization, let us challenge the bold assertion that we made at the beginning of this chapter, that the moon will be the first place in space that humans will colonize. Though this will almost certainly be true for the list of natural celestial objects, we must also ask, What about constructing human habitats in space, either much closer to Earth than the moon or at roughly the moon's distance from our planet? Don't these "space habitats" offer superior advantages and a greater ease of construction in comparison to colonizing the moon?

Before we analyze these questions, we must deal with a semantic issue. Any habitat in space that orbits only a few hundred miles above Earth, like the Space Station on which construction has finally begun, should not be called a space colony. Because of the ease with which humans can pass from Earth to the habitat and back again, on a journey lasting only a few hours, it will not be inhabited by people who plan to devote most of their lives to living in space. Instead, such habitats will see technicians, scientists, and other specialized workers come and go at regular intervals, probably spending no more than a year or two in orbit around Earth. For most people, these interludes do not amount to the colonization of space.

The larger possibility—that we could build giant space habitats far from Earth, human-made miniworlds that might each hold tens of thousands of colonists—seems to be an idea whose time has come and gone. During the 1970s, the physicist Gerard O'Neill envisioned enormous, rotating cylinders, many miles long and several miles across, made of material mined from the moon and flung into space, as homes for tens of thousands of people who would live on the cylinders' inner surfaces, colonizing new environments to surmount the growing problems of an overcrowded Earth. The cylinders would use solar power to grow crops, to process raw materials, and to maintain a technological infrastructure; their rotation would provide a simulation of Earth's gravity through what is familiarly called centrifugal force. To achieve stable orbits and to use lunar raw materials with relative ease, the cylinders would orbit Earth at the moon's distance. Some day, O'Neill and his associates predicted, these space habitats could have a population numbering in the millions, a generation not lost in space but born and raised there, with new and quite different ideas about what it means to be human.

O'Neill's vision now seems to be a reflection of the worries of past decades, culminating in the belief that life on Earth threatens to become so unbearable that a lucky few must attempt to escape from it. Even though one can cogently argue that overall conditions have in fact grown worse on Earth, our attitudes have shifted, perhaps with the ending of the cold war, so that almost everyone now sees us as here for the long haul, destined to confront Earth's problems here on our own planet even if we succeed in colonizing other objects in the solar system. The concept of mining minerals from the moon, launching them into space, and assembling them into giant artificial colonies seems far less promising, at least in the near term, than using the moon itself as both a home and a source of raw materials. A devoted follower of O'Neill might simply say that a lunar colony may well come first, but eventually the lure of creating entire new miniworlds will prove stronger than the wish to live on existing objects. We can reasonably postpone this debate until a time when at least one lunar colony has been well established.

On the other hand, when we look to the far future, we can consider an advanced civilization that extends O'Neill's dreams to the point that it encloses its star in a shell of colonies, capturing all of the star's light and heat for its own purposes. In the solar system, natural objects lie at such great distances from the sun that nearly all of the sun's output radiates into deep space. Earth, for example, intercepts only about one part in a billion of the energy that the sun produces each second. We can imagine a society, millions of years in the future, that has become capable of taking apart the sun's giant planets and turning their raw materials into space habitats. These habitats could intercept essentially all of the sun's power, employing it for human purposes. In that case, instead of the six billion people that Earth now supports, we could envision a human population roughly one billion times greater, all deriving their energy from a single star.

Among space enthusiasts, this scenario has the name "Dyson sphere," a reminder that Freeman Dyson first proposed the idea that an advanced civilization might find a way to use a star's entire energy output rather than letting it escape. Dyson noted that if this should occur, we would expect not to see the star at all, since its light would be blocked by the habitats around it. Instead, we might look for the infrared (heat) radiation emitted by the civilization—a civilization so far in advance of ours that Earth's enormous human population might number only one billionth of theirs.

Before we sail into these uncharted realms of possible advanced civilizations, we must refine our own to the point that we have some confidence in its long-term survival. Those who advocate colonizing the moon and Mars see these efforts as the key to human progress, a new frontier mentality that will encourage innovation. Who will direct these changes? How will society be governed once humans leave the Earth for other celestial abodes? Let us turn to the no longer mundane topic of human governance and ask ourselves, Who will rule in the heavens?

Chapter 4

The Sociology of Space

During the nearly four decades that have elapsed since astronauts first entered Earth's orbit, humans have surpassed their early records for time and distance only modestly. Humanity's most distant voyages through space have occurred on eight *Apollo* missions to or around the moon, each involving a round trip of half a million miles, while the flight of longest duration in Earth orbit lasted for less than sixteen months—still an impressive feat for a Soviet cosmonaut provided with little in the way of comfort or entertainment. The largest gatherings in space occurred in 1995 when American and Soviet astronauts docked the Space Shuttle with the *Mir* spacecraft for a few days; thus ten people orbited the Earth while technically sharing the same living space.

Unless a pall descends on our desire to explore Earth's cosmic neighborhood, these modest records from the second half of the twentieth century, when space travel began, are destined to be broken during the first half of the next century, which is likely to see humans spend years in a spaceborne environment and even journey to Mars. If we establish a lunar colony, some of its inhabitants may spend the bulk of their lives there and before long could be raising moon-based families. By the end of the twenty-first century, we should be contemplating journeys to the outer solar system, especially to the two largest planets, Jupiter and Saturn, and their satellites; even with improved space propulsion, these voyages might well take decades for a round trip.

Who Will Govern Space?

The twenty-first century should therefore bring us to an era in which many dozens or even many hundreds of people live for years in extraterrestrial environments, and should usher in an epoch in which these numbers grow into the thousands and beyond. Along with our speculation about the technology of these future operations and about the targets of exploration and colonization toward which future astronauts will direct their efforts, we can justifiably ask, Who will go into space? Who will pay for space operations? Who will choose and enforce the laws governing human activities in space? Can we reasonably expect to evolve new rules on a sufficiently rapid time scale to keep up with the technological progress we might quickly achieve?

Who Will Choose the Space Colonists?

Before any dreams of creating new societies in space can even begin to take shape, someone must resolve the issue of who will build the new societies. Historically, individual countries have first sent members of their armed forces into space, followed by scientists, engineers, and others who could take the greatest advantage of their experiences. Politicians have also orbited the

Earth, on the theory that they could help to promote space exploration on the ground. During the 1980s, NASA decided to send ordinary civilians into space, qualified only by their eagerness to ride the Space Shuttle and their willingness to share their impressions with the public, and chose a New Hampshire schoolteacher, Christa McAuliffe, as the first civilian representative. Unfortunately, McAuliffe died, along with six other NASA astronauts, when the Space Shuttle *Challenger* exploded shortly after liftoff in January 1986. Since then, NASA has kept on hold any plans to send ordinary citizens, or even journalists, into space.

Even if we put aside, as a relic of cold war attitudes, any competition to establish a military presence in space, we must recognize that no society is likely of its own volition to create a new society based on principles fundamentally different from its own. The urge to experiment does not extend that far; instead, we can expect that if the United States government establishes a colony on the moon, the new colonists will be expected to follow the same laws that govern the United States, suitably modified for another celestial object. If we attempt to raise the level of our dreams and imagine that the United Nations will create a new constitution for a lunar colony, we shall immediately see that disagreement would arise concerning the population of the colony and the governance of its members.

What if space colonies are created not by governments but by groups of individuals? This sounds promising to those who favor a diversity of approaches, but the dream quickly runs into the reality that colonizing space costs money. One can hardly expect giant corporations to create colonies of anarchists, any more than a capitalistic society would create a new variant of communism to see whether it might for once function in the manner that Marx imagined.

Who Will Pay for Space Exploration?

Under the familiar principle that the piper plays what the payer chooses, the question of who will go into space takes us quickly to the question of who will pay for their voyages. Past experience, the impressively large costs involved in space exploration, and the fact that economically viable returns will lie many years in the future combine to suggest that only a government, and perhaps no single government at that, can afford to invest in space colonies. Of all human activities, the colonization of space stands out as the most suited to international collaboration. Or is this the discredited view of outmoded liberals who fail to see the power of private enterprise?

Until now, all successful efforts at space exploration have resulted from governmental activities, as the United States, the Soviet Union (and now Russia), Japan, and France have all built and launched Earth-orbiting space vehicles. With the end of the cold war, the trend toward international cooperation in space has strengthened, with the United States's NASA in the vanguard (at least in spending), Europe's ESA not far behind, and Japan's ISAS in third place. But can we expect that NASA and ESA will be charged with the long-term colonization of space simply because they have acquired expertise in space exploration largely with automated spacecraft? The advent of space colonization, at least as a serious possibility, could herald a key change in how humans "run" space. A supremely rational approach might see the creation of an international organization, or a unit of the United Nations, to direct such activities. After all, when we imagine other societies that might be sending teams of explorers into space, we almost inevitably conceive of them as acting as a single political entity and never dream of meeting the "American" or "European" equivalents of the team from a particular planet such as Vega 12. When shall we and when should we cease to explore the cosmos in our national capacities and begin to do so simply as representatives of the human race?

One answer to this question, not favored by old-line liberals, sees private efforts at space exploration as a natural way to achieve the internationalization of space. Acting on the most fundamental of human impulses, the urge to profit from nature's bounty, entrepreneurs are fully prepared to ignore, as far as possible, the limitations imposed by national boundaries. The notion that international corporations lead to international cooperation, rather than follow it, could reach its full flowering in space. In the short term, however, what we might see is a competition between nationally (and internationally) funded governmental operations on the one hand and private entrepreneurs on the other.

The United Nations Treaty on Outer Space

For those who seek to coordinate human efforts to explore celestial realms beyond the Earth, the United Nations offers the most obvious mechanism. Created in 1945 in response to the ravages of World War II, the United Nations has proven highly useful in providing a forum for opposing opinions, though it has been far less effective as a means of preventing or ending military encounters between armed states. The latter failure arises from the fact that because the United Nations has no enforcement powers of its own, it must rely, as perhaps it should, on mutual agreements among its member states (or at least among the members of its Security Council) to impose its will—typically with armed soldiers—in the pursuit of peace. Should Earth's political clashes find an echo in outer space, the military weakness of the United Nations would become even more apparent. The most effective means that the United Nations can offer to prevent armed encounters in space lies in the opportunity it offers for potential enemies to thrash out their differences here on Earth.

This fact has dominated the successful efforts of diplomats to create international treaties to curb military activities in space. During the 1960s, the United States and the Soviet Union launched satellites by the hundreds, many with important military applications, and competed for what military minds often called the "high ground" of the moon. Influenced by their experience with artillery on Earth, even great generals tended to lose sight of the fact that because the moon's gravity must be overcome before a rocket can start toward Earth, launching a rocket to attack Earth from the moon will always cost more energy, take much longer, and require more complex guidance than launching the same rocket from an orbit around Earth. The fact that a moon-launched rocket must travel a far greater distance than one in orbit only a few hundred or a few thousand miles above Earth's surface provides a greatly increased opportunity for the enemy to intercept and destroy the rocket before it reaches its target.

In any case, international diplomacy created a stunning success—on paper. In October 1967, on the tenth anniversary of the launching of the first artificial Earth satellite, the United Nations announced its Treaty on Principles Governing the Activities of States in the Exploration and Use of Outer Space, Including the Moon and Other Celestial Bodies, called the Outer Space Treaty for short. This treaty, recognizing "the common interest of all mankind in the progress of the exploration and use of outer space for peaceful purposes," states that "the exploration and use of outer space, including the moon and other celestial bodies, shall be carried out for the benefit and in the interests of all countries, irrespective of their degree of economic or scientific development, and shall be the province of all mankind." Article II of the Outer Space Treaty agrees that outer space is not subject to claims of national sovereignty; Article III states that signatories must abide by international law and must avoid contaminating the celestial objects they visit; while Article IX requires that states' facilities be open to visits by other states on a reciprocal basis. Most important at the time, Article IV of the Outer Space Treaty forbids the placement of nuclear weapons in orbit around Earth or anywhere else in space and states that "the moon and other celestial bodies shall be used . . . exclusively for peaceful purposes," with all military installations and maneuvers forbidden although military personnel may participate in peaceful space exploration and scientific research.

This has been a good set of principles to follow. Cold war tensions throughout the two decades after 1967 left many military and political leaders on both sides of the conflict eager to place weapons of mass destruction in orbit around Earth, reducing the time required for them to reach their targets. The Outer Space Treaty made it impossible to put hydrogen bombs in orbit openly and highlighted the dangers of a race to do so covertly. Such a race would likely end with the revelation that both sides had violated the treaty and leave the world even more fearful of weapons of mass destruction—a result that would reflect negatively on both sides of the cold war conflict. More significant, it seems, was the fact that the Outer Space Treaty gave both sides the chance to avoid a secret race to put nuclear weapons in orbit while looking properly noble in their refusal to do so.

Thirty years later, the Outer Space Treaty remains the United Nations' main achievement in promoting peaceful space exploration. Further treaties followed, including an agreement in 1968 on the rescue of astronauts, an agreement in 1972 on payment for damages caused by artificial objects that strike the Earth, and a convention in 1975 dealing with the registration of objects launched into space.

The Moon Treaty of 1979

In 1979, however, the treaty-making process encountered the growing sentiment that human desire to explore the solar system solely for scientific purposes should give ground to a need to derive wealth from space. Long negotiations (understandably, there are no short ones at the United Nations) produced the Agreement Governing the Activities of States on the Moon and Other Celestial Bodies, informally known as the Moon Treaty to distinguish it from the Outer Space Treaty of 1967 and also because it centered on activities that governments might carry out on the moon. The Moon Treaty attempted to specify the types of activities either allowed or prohibited by the general language of the Outer Space Treaty. Article 4 of the Moon Treaty states that "the exploration and use of the moon shall be the province of all mankind and shall be carried out for the benefit and in the interests of all countries, irrespective of their degree of economic or scientific development. Due regard shall be paid to the interests of present and future generations as well as to the need to promote higher standards of living and conditions of economic and social progress and development in accordance with the Charter of the United Nations." Article 6 states in part that "in carrying out scientific investigations . . . the States Parties shall have the right to collect on and remove from the moon samples of its mineral and other substances." Article 7 requires that "in exploring and using the moon, States Parties shall take measures to prevent the disruption of the existing balance of its environment" either by contamination or by changing the lunar environment and shall also avoid harming Earth with extraterrestrial matter. Most significant in terms of future governance of a lunar colony, Section 5 of Article 11 of the Moon Treaty states that the signatories "hereby undertake to establish an international regime, including appropriate procedures, to govern the exploitation of the natural resources of the moon as such exploitation is about to become feasible." An additional article provides for review of the Moon Treaty after ten years, and for a conference after five years to review the implementation of international regulation of lunar activities.

Though heartily ignored by the general public, the Moon Treaty provoked a spirited lobbying campaign by space advocacy groups who correctly noted that the treaty would limit or prohibit activities such as lunar mining. These lobbying efforts culminated in success: The United States Senate refused to ratify the treaty, and the Soviet Union, though mystified as to the reasons behind this action, promptly likewise rejected it. The Moon Treaty thus represents the most far-reaching attempt to impose an international governmental regime on the moon (and, by implication, on other celestial objects)—an attempt that was rejected first and foremost by the United States at the behest of its citizens (at least of those who effectively made their opinions known).

The successfully ratified Outer Space Treaty and the rejected Moon Treaty share a common attribute. They were negotiated by governments to deal with governmental actions. Of course, each government makes itself responsible for the actions of its citizens, but the thrust of these treaties implies that governments will rule space, either directly or through their common creation, the United Nations.

We now face a different situation. As individuals, often in corporate guise, begin to participate in activities beyond the Earth, agreements such as the Outer Space Treaty and the Moon Treaty lose much of their relevance. Under the existing hierarchy of power on Earth, each government regulates its own citizens. But if a corporation plans to launch a satellite or to send an expedition to mine the asteroids, what will the United Nations say? Under the existing structure, essentially nothing. What will the corporation's government say? Today, large corporations can escape most governmental control by the simple step of finding another, more malleable government. Does this mean that space represents the final frontier in law as well as in distance? To some extent, it does. As we shall see in Chapters 5 and 6, bold visionaries see space as a place to assert their rights of ownership against all comers—including governments that once imagined that they would rule space. In the not-so-long run, this approach is doomed to chaos and failure. For now, the "privatization of space" emphasizes the difficulties that human society will encounter as it opens new environments with only old laws for a single celestial object to guide it.

Who Will Write the Laws for Space Dwellers?

More fascinating than the question of who should and will govern space, because they are closer to our daily lives, are the questions of how those living in space will be governed and who will be chosen to be governed. Simple extrapolation from the situation on Earth suggests that "space law" has a chance to expand without limits. For decades attorneys have discussed questions concerning who has jurisdiction over crimes committed in space and which forum is the most appropriate for trying civil suits arising from alleged spaceborne torts. Among the most interested participants in these debates have been retired naval officers, who stress the analogy between a spacecraft and a ship on the high seas, where the captain's word is the law. This comparison suggests that spacecraft should be governed by a set of regulations, which creates a hierarchy of authority roughly similar to that of a military unit. The commander on the scene would have the authority to decide cases, and these decisions would be appealable to a higher authority once the spacecraft returns to base.

This analogy has some merit for short excursions into space or across the oceans with a foreseeable return to Earth or into port. The classical roots of the word "astronaut" describe a "space sailor" and imply a person on a journey far from home but connected to that home by custom and law. To the extent that the activities of astronauts represent a temporary departure from the norm, it makes sense to apply a set of rules different from those that govern society at large. But what should and will occur once people begin to live in space indefinitely? Does it make sense to bring their laws with them, no doubt with various adaptations, or instead to create a new set of laws, a "space code" to incorporate the different facets of life on our teeming planet and life within a small population in an entirely different environment?

Posing these questions, whose answers may determine how eagerly and how effectively humans become space dwellers, leads straight to an ancient issue, the question of how an ideal society should be organized. From biblical lawgivers to Plato and Muhammad, from Cromwell to Locke, from Jefferson to Lincoln and so to the present, humanity has never lacked for those who claim special insight into how we ought to live, or for experiments that attempt to obey at least some of the commandments laid down by philosophers and politicians. Most of the difficulties in following these precepts have arisen from the multifaceted nature of human personality, multiplied by a wide diversity of individual backgrounds and interests. Among the attempts to reconcile individual freedom with social order, western democracy seems best—to those who enjoy its fruits. But is it right for space? Should we not instead make a conscious effort to rethink our standards for society, recasting them in a form suitable for a small colony of explorers in space who in turn must reshape them as their colonies become more mature and self-sustaining?

These provocative questions should gain increased prominence, once significant numbers of people begin to live in space. In one sense, "should" represents only a pious hope, a belief that the social framework deserves periodic reinspection. We can reasonably anticipate that for centuries to come, if not for millennia beyond, the portion of the human population that lives away from Earth will remain minuscule. Why, then, should the majority care about what happens to this modest fraction when Earth-based problems will continue to demand all the immediate and serious attention we can provide? Why should the organization of space colonies attract us more than the problems of Uruguay or Zimbabwe?

The answer surely lies in the human capacity to dream, coupled with the fact that colonies in space offer virgin territory in both the physical and moral senses of the term. With space colonies we have a chance to make a fresh start. In fact, we might envision a situation, described more fully in Chapter 3, in which repeated fresh starts can be made in different space colonies, each of them an artificially created world with its own rules and consequently its own sort of inhabitants. On the other hand, it also seems likely that the first set of rules, adopted for the first locale where humans see themselves as permanent inhabitants, will have a strong effect on future generations. For this reason, it is worth asking beforehand what will or should be the appropriate set of regulations to govern, say, a colony on the moon or a base on Mars. Though the astronauts who create these new habitats will begin by imagining them as extensions of

Earth, before long they will see themselves as residents of a new city on a hill, capable of creating a rule of law that will reflect both their human heritage and the new environment in which they live.

New Rules for New Situations

When we turn our attention from colonies to be established within the solar system and speculate about the rules that would govern a group of humans sent on an interstellar journey, two aspects of our speculation seem to come into focus. First, any such journey would require many human lifetimes, and so the colony would effectively establish its rules on the journey long before actually reaching its goal, or else the voyage to the stars would have to depend on technology far from its final development (see Chapters 9 and 10). In the former case, it is clear that the colonists would indeed form a world to themselves, subject to terrestrial oversight in the sense that messages could be exchanged with a government on Earth but effectively free of terrestrial commands in terms of practical enforcement. In the latter case (or in the former, once the destination has been reached), any significant exchange of messages would occur only over distances of many light-years. The fact that the speed of light represents a maximum implies that long intervals would elapse between any message sent to or from Earth and its contact at the other end of an interstellar distance. This factor alone would provide a strong disincentive to respond to Earth-based commands.

Will Space Colonies Be Independent and Self-Regulating?

In addition to any formulistic regulations and laws that attempt to govern the behavior of individuals, the most significant checks and guides for an individual's behavior within a group of humans are the attitudes that the group brings to bear on an individual and the extent to which these attitudes are internalized by the individual. In the distant past, the relevant human groups numbered no more than a few dozen or a few hundred, and social governance arose from and consisted of essentially personal interactions. As societies grew larger, new techniques of social influence and control had to be, and promptly were, developed, and eventually they were extended to the point at which societies with millions of individuals could be encompassed within the same basic framework, an achievement often referred to as the rise of civilization.

To escape civilization will be the conscious or unconscious goal of many of those who seek to join colonies in space, which for the indefinite future are destined to include relatively small numbers of people with an obvious requirement for a more flexible structure of governance. The closest analogy in our history is the expeditions to the Earth's polar regions early in the twentieth century, especially those that spent a year or more in Antarctica. Perhaps the best example consists of the group of twenty-six men led by Ernest Shackleton, who left England in 1914 with the intention of crossing Antarctica for the first time but found themselves trapped in the pack ice of the Weddell Sea, which eventually destroyed their ship and left them drifting on the ice. The story of how these men managed to reach a desolate island in three small boats, how Shackleton and five others took one of these boats over nine hundred miles of the most fearsome ocean on Earth to reach a whaling station on South Georgia, and how he rescued all of his crew the next year provides a heroic story of human endurance. It also demonstrates how these men, utterly isolated for more than two years from a world that in fact had written them off as lost forever, maintained their unity and naval discipline. Those who have analyzed their behavior conclude that the men survived because they put their trust in Shackleton, who had a rare and uncanny ability to judge his men and their failings, both real and potential.

But what would have happened to the twenty men on the island had Shackleton's small boat been lost at sea? His second-in-command, Frank Wild, was a fine leader, but no Shackleton; in addition, the morale of the group understandably declined as month after month passed with no sign of rescue. Even though these survivors could have continued to live on the meat of seals and penguins, it seems likely that their survival as a society would not have been possible if they had been forced to pass a second winter on their desolate island, although one should not underestimate how tough men were in those days. Today, a group of men and women, likewise numbering a few dozen, spends the southern winter at the South Pole station, maintaining radio contact with the outside world but unreachable by aircraft except for an airdrop halfway through their period of isolation. Despite the fact that this group can exchange messages with the outside world and receive television broadcasts, its members, quite understandably, adopt a set of behaviors quite different from those that they display "off the ice." These behaviors can be roughly described as a regression to childlike attitudes, including a "we won't clean up until our parents come home" approach to housekeeping.

Like Shackleton's expedition, the South Pole "winter-overs" have a leader who, like Shackleton, must judge carefully how to guide the group through a difficult period. But, like Shackleton and his men, the winter-overs rightly believe that their time in the wild is temporary, with a return to civilization only months in the future. What would it be like if a Shackleton led an expedition designed *not* to return to its point of origin? Would the men and women who went with him exhibit the same striking loyalty that the original Shackleton commanded?

The answer seems to be yes—if a Shackleton could be found. The men on the 1914–16 Antarctic expedition responded to Shackleton not because he carried the authority of a government or a large social group but because he demonstrated his ability to solve problems on the spot in a multitude of situations where survival depended on finding the correct solution. If Shackleton were instead the captain of a spaceship bound for, say, Jupiter's moon Ganymede with the intention of founding a human colony there, those who journeyed with him would respond in a similar manner.

Making the Transition from a Charismatic Leader to a Sustained Society

A space colony's more difficult social task would be to find and to follow rules that do not stem from a single person's directives but instead allow the group to function successfully without a leader of Shackleton's talent. This problem mirrors that of developing societies, often led by one or more persons with exceptional skills. With the passing of the founding fathers, either charismatic individuals attempt to seize the mantle of leadership or the society finds a way to institutionalize what the first leaders achieved by force of personality. The third alternative, of course, is failure of the group to survive. The history of the United States reflects the second possibility, while highlighting the difficulty of this transition. The country paid dearly for the (perhaps necessary) confusion produced by the lack of leadership displayed by presidents such as Franklin Pierce and James Buchanan, which left prominent senators and their factions free to engage in the struggles that culminated in civil war. On the other hand, countries in which the first, autocratic alternative has been instituted demonstrate that sooner or later this approach will fail to produce a sufficiently charismatic and successful leader. For every Marcus Aurelius there is a Commodus, corrupt, inept, and destructive. The unwieldy size of even modest countries magnifies this simple truth, but it has proven true for much smaller groups, such as the Mormon community in Utah after Brigham Young died.

How will the first space colonists make the sort of transition that the Mormons did, from an autocracy in which a single individual's word was law to a far more democratic situation in which a council of elders established religious doctrine but other issues are decided by a democratically elected legislature? The Mormons had the advantage of living within the United States, which was increasingly in a position to enforce its laws as the West became more thickly settled. For a space colony within the solar system, this sort of assistance would arise from the exchange of radio and television messages, with the possibility that emissaries from Earth could arrive within a few months' time. This situation can be compared with the Roman Empire, in which a governor effectively ruled the more isolated colonies, such as Spain and Britain, with the possibility that the central power at Rome might in extreme cases overrule or dismiss that governor.

When we look to colonies far beyond the sun's planetary realm, so far from Earth that many years must pass for even an exchange of messages, we enter uncharted social territory. Science fiction novels have speculated on the changes that society must undergo to allow new regimes of behavior and governance in such colonies; certainly past eras on Earth furnish little guidance. As the twenty-first century opens, we are about to test these mental projections against reality, for we stand on the brink of establishing colonies on the world most favored by astronomical speculation, the red planet Mars.

Chapter 5

When Babies Are Born on Mars

Throughout Earth's history, the reddish disk of the planet Mars has punctuated the skies of Earth, drawing admiration for its color and its strange wanderings through the heavens. Like the other planets, Mars continuously changes its position in the sky with respect to the "fixed stars," which owe this ancient designation to the fact that they appear to maintain a rigid arrangement in the starry constellations. The Greek word *planet,* which means "wanderer," describes the observation that the planets seem to sail among the stars, though they always remain within the band of constellations called the zodiac. The planets' ever-changing positions provided thought-provoking material for those who sought to explain the motions of the heavens, because they created grave problems for the concept of a single "celestial sphere" that turned once each day around Earth, carrying with it the sun, moon, planets, and stars.

After a few dozen centuries of brilliantly misguided imagining, astronomers eventually realized that the planets are worlds like Earth, orbiting the far more massive sun. To achieve this insight, astronomers had to abandon the intuitively obvious notion that the Earth occupies the center of the cosmos, forever fixed in place while the vault of heaven rotates around it. During the sixteenth and seventeenth centuries, the news that Earth is a planet spread from astronomical circles to the rest of society, producing considerable spiritual unrest. Even today, when every educated person learns that we live on a planet orbiting a star called the sun, not all of us fully internalize this concept. Indeed, we may ask, Why should we? The belief that Earth stands still, central and motionless, satisfies the soul and fulfills our intuitive feelings. The contrary notion, that we inhabit a mote of dust with no greater claim to attention than perhaps countless other planets orbiting distant stars, offered little or no payoff.

Now, however, the opening of the third millennium brings with it the promise of humans colonizing other objects in the solar system, not just the modest satellite that orbits our own planet but full-fledged planets comparable in size to Earth and no doubt possessed of enormous mineral wealth. Like Columbus, we can dream of worlds whose riches beckon to us, this time vowing to do things right. Of all the new Indies that beckon us outward, urging us to leave our terrestrial ways to establish advanced colonies throughout the solar system, one world, rust-red Mars, ranks at the top of the list.

Mars Attracts Us

Mars is not the planet that comes closest to Earth (that distinction, by a few million miles, belongs to Venus), nor the largest planet in the solar system (Jupiter and Saturn rank first and second with eleven and nine times Earth's diameter, respectively, while Mars ranks only seventh), nor even the largest of Earth's closest neighbors (Venus, almost as large as Earth, wins that prize). What Mars presents to us, unlike any of the sun's other planets, is a world enveloped in an atmosphere something like our own, with a visible solid surface whose features change with the Martian seasons. Only half as large as Earth in diameter, Mars has a total surface area barely more than one-quarter of Earth's; on the other hand, since Mars has no oceans, its "land" surface roughly equals that on Earth. The reddish coloration of this land can deservedly be called rusty, since the iron in Mars's soil, following the same process that made the glorious, rust-red colors of the rocks lining the Grand Canyon in Arizona, is slowly oxidizing, that is, combining with oxygen in the atmosphere.

Oxygen in the atmosphere! We all breathe oxygen and are eager to find it on other planets. Does the oxygen on Mars make it ready for human habitation? Not really, because the oxygen is nearly nonexistent. Oxygen has a total abundance in the Martian atmosphere less than one ten-thousandth of its abundance in Earth's atmosphere. Mars has an extremely thin atmosphere capable of producing only weak atmospheric pressure at the planet's surface, less than 1 percent of the surface atmospheric pressure on Earth. The atmosphere of Mars consists mainly of carbon dioxide, which, along with the inert gas argon, forms more than 99 percent of the atmospheric total. Only trace amounts of oxygen and water vapor exist in the Martian atmosphere—sufficient to "rust" its rocks over billions of years but far too little to be of much use to any explorers visiting the red planet. The small amount of water vapor comes from the Martian polar caps which contain large amounts of frozen carbon dioxide ("dry ice" in familiar speech) along with frozen water (ordinary ice). The tiny amount of atmospheric oxygen, even less than the water vapor abundance, arises when sunlight breaks apart some of the water vapor molecules, producing atoms of hydrogen and oxygen.

So the atmospheric glass on Mars seems considerably more than half empty. On the other hand, if we are looking for compounds that are familiar and useful to us, the Martian surface is more than half full. Not only does Martian soil consist of iron-rich compounds that might be mined to provide construction materials, but it also contains significant amounts of water, mixed into the rocks and frozen solid. Like the permafrost in northern Siberia and Alaska, Martian soil never melts into slush, but if it were heated artificially, it could be "mined" for its water as well as its mineral content.

Thus the surface of Mars apparently offers abundant supplies of raw materials for potential colonists. Before we consider them, we should mention the present "colonists" of Mars—the Martians who have fascinated humans for more than a century. Does Mars harbor life? Why shouldn't it? If Mars does have life, or did so in the distant past, doesn't this demonstrate how ripe it is for colonization?

The possibility of life on Mars has been good for astronomy; one can hardly name a topic more appealing to the public nor one that can provoke greater attention and controversy. During past centuries, astronomers who studied Mars with telescopes saw surface features which allowed them to deduce that the planet rotates in 24.5 hours, just a bit longer than the Earth's rotation period. They also noted the presence of polar caps, which came and went with the Martian seasons. Like Earth, Mars has a rotation axis that tilts by about twenty-four degrees from being perpendicular to the planet's orbital plane. As Mars orbits the sun, each hemisphere in turn receives sunlight more directly during summer and then undergoes winter as sunlight shines more directly on the other hemisphere. Since Mars takes 1.8 years to orbit the sun, each season lasts almost twice as long as it does on Earth, but the same effect occurs on both planets.

Water, the Bringer of Life

As the seasons change on Mars, so too do some of the blotchy patches that darken its surface. Could these changing surface features be Martian vegetation made to flourish by the arrival of spring water? In 1877 the Italian astronomer Giovanni Schiaparelli reported that he had observed "canali" on Mars. With this word, Schiaparelli denoted straight-line features; the Italian word does not denote a water-directing trench. Soon after this announcement, a rich, well-born American, Percival Lowell, became fascinated by Mars. Lowell thereafter devoted much of his fortune to the construction of an observatory near Flagstaff, Arizona, to observe the red planet and announced that he had seen canals on Mars. And not just a few canals—Lowell had seen canals by the dozens, some of them double and triple, and all of them connecting what he called "oases." Lowell drew maps of Mars and its canals from which only a dunce could fail to conclude the existence of a planetwide network of canals; only a small leap of the imagination, which Lowell happily furnished, was needed to suggest that a Martian civilization had built this network to carry water from the polar vicinities to the warmer, water-poor equatorial regions of the planet.

Lowell's Mars, with its water-starved inhabitants presumed to be members of an advanced, degenerate race, fascinated the public a century ago, the more so after H. G. Wells wrote *War of the Worlds*, a novel about an attack on Earth by aggressive Martians. Martian canals remained controversial for five decades more until spacecraft finally sailed by the planet and photographed a desiccated surface, devoid of canals or of anything else suggesting the existence of a technologically advanced civilization.

We now know that in fact liquid water simply cannot exist on Mars for the simple yet sufficient reason that the atmospheric pressure is too low. A molecular compound—water, for example, or carbon dioxide, for another—can exist in the liquid state only if atmospheric pressure keeps a "lid" on the liquid; if it does not, the compound will vaporize when heated, or solidify when cooled, passing directly from the solid to the gaseous state or back again. Carbon dioxide behaves this way on Earth: A solid block of dry ice sublimates (as chemists call it) directly into carbon dioxide gas without leaving a pool of carbon dioxide—a fact that mystery writers have employed to hang a victim slowly as a large piece of dry ice sublimates, allowing gravity do its work. On Mars, with a surface atmospheric pressure just 0.6 percent of Earth's, neither carbon dioxide nor water can exist as a liquid. This is bad news for those who love life, for life in anything like the form we know seems to require a liquid solvent, a fluid in which complex molecules can float and gently interact with one another.

The prospect of life on Mars therefore appears bleak, though we cannot rule out forms of life that have evolved to survive by profiting from the momentary appearance of tiny amounts of water, like organisms that live within the

Siberian permafrost. The term "evolve" seems entirely relevant because evidence strongly indicates that liquid water once *did* flow in quantity over the surface of Mars. This evidence consists of large numbers of channels and rivulets, almost certainly carved by running water, first photographed from the two *Viking* spacecraft that orbited Mars for more than a year during the late 1970s. The number of craters left in these ancient streambeds suggests that water last flowed on Mars several billion years ago. The conclusion follows that since that time Mars has lost a significant part of its atmosphere and with it the ability to maintain any liquid water on its surface.

Viking Searches for Life on Mars

The *Viking* mission to Mars sent two landers to the surface where they dug samples from the soil and dropped them into automated laboratories designed to test for the presence of microorganisms. Although the initial results seemed positive, more lengthy, thoughtful analysis led nearly all the *Viking* scientists to conclude that chemical reactions on Mars had mimicked some of what microorganisms do on Earth. Coupled with the absence of any sign of life on Mars's dusty surface (which is raked by winds that reach hurricane-like velocities but in the thin atmosphere have far less effect than strong winds on Earth do), the *Viking* results led to two decades of negativity concerning the chances of finding indigenous life, or even signs of long-vanished life, on what seemed to be the last, best hope for extraterrestrial life in the solar system. During the 1980s and early 1990s, Earth seemed likely to be the only living planet to be found for a long, long distance and a correspondingly long time. Belief in life on Mars, stoked by Lowell and Wells and long a cornerstone of space exploration motivation, became largely the property of those who insisted that the *Viking* orbiters had photographed a mile-wide face, carved by a vanished race of intelligent and anthropomorphic Martians whose true nature NASA and our government leaders concealed from the populace.

Could Life Exist on Mars Now?

These attitudes changed markedly within only a few years. First, Jupiter's moon, Europa, and Saturn's moon, Titan, turned out to offer better prospects for life than had been previously thought (see the next chapter). Second, during the latter half of the 1990s, astronomers made the first discoveries of planets around other stars, creating a list that runs well past a dozen (so that we now know of more planets orbiting other stars than orbiting the sun) and placing further discoveries of "extrasolar planets" in the ho-hum category. Third, in August 1996, NASA announced that scientists had discovered signs of ancient life in a meteorite from Mars that had been found in Antarctica. Coupled with the success of the *Mars Pathfinder* mission in the summer of 1997 and with the ongoing *Mars Global Surveyor* mission, the "Mars rock" brought the possibility of life on Mars back into public attention, pointing the way toward future missions to the red planet that not only could resolve whether Mars had life billions of years ago but might also determine whether life survives on Mars even today, perhaps pushed to the edge of extinction but stubbornly holding on beneath the arid surface of the red planet. *Mars Global Surveyor* also performed the useful function of photographing the "face on Mars" in detail, revealing it to be a natural rock formation—unless, of course, NASA or the Martians, having glimpsed what pain this formation might provoke, had quickly dismantled it.

If we believe that life may exist on Mars, how should we look for it? Wells's vision of waiting for Mars's warriors to attack Earth has the merit of simplicity (see Chapter 11), and Lowell's approach of deducing the existence of Martians from careful telescopic observation likewise requires only modest efforts. The dozen meteorites known to have come from Mars, identified as such because their chemical composition precisely matches that measured on Mars by the *Viking* landers and differs significantly from that of all rocks on Earth, also offer a relatively inexpensive means of examining the Martian surface, but one that has so far proven inconclusive. The indications of fossil life, which seemed so promising at first sight in the summer of 1996, now appear more likely to be the result of overzealous interpretation, plus possible contamination by terrestrial microorganisms during the thousands of years that ALH 84001, the famed Mars rock, lay on the Antarctic ice. Nevertheless, the fact that scientists could find such highly suggestive evidence in even one Mars rock among a mere dozen, and indeed in the only truly old rock in the collection (all the others have ages of less than about a billion years, whereas ALH 84001 clocks in at 4.3 billion years, older than any rock on Earth, and provides a relic of the time when water flowed copiously on the Martian surface) definitely whets the appetite of astrobiologists for further investigations on Mars.

The Mars rock still has its adherents, who argue that its ancient evidence speaks strongly of the existence of life, as well as a larger number of skeptics, who feel that this evidence falls far short of compelling. All agree, however, that we should seek further evidence if we hope to resolve this gripping issue. Finding more ancient meteorites from Mars may require many years; in any case, we cannot expect to examine the full range of Mars's history in this manner. To search for possible ancient life on Mars or for life that persists even today, we must take the obvious necessary steps and return to the red planet.

Future Exploration of Mars by Automated Spacecraft

During the first decade of the twenty-first century, automated spacecraft should explore Mars far more thoroughly than the best efforts of the 1990s. Three such explorers, one Japanese and two from NASA, are now sailing toward Mars and are destined to arrive as the year 2000 begins. Two of these spacecraft will enter orbits around the planet, making detailed observations of its surface from altitudes of a few hundred miles, while the third, NASA's *Mars Polar Lander,* will make humanity's first contact with the ground close to one of Mars's polar caps.

Using a giant parachute to brake its descent through the thin atmosphere, the *Polar Lander* will land about five hundred miles from the south pole of Mars. Five minutes before this landing, the spacecraft will release two sturdy probes which will fall onto the surface. Each probe will excavate a small crater, punch through the crater bottom for another few feet, drill into the soil, and use laser beams to heat some of the dirt and grit that it digs up. This heating will vaporize the lighter compounds in the soil, which the instruments on each probe can measure to determine how much water and carbon dioxide has been trapped within the soil below the surface. The *Polar Lander* itself will use a robot scoop to gather Martian soil closer to the surface, which it will bake in an oven, again releasing volatile compounds that have been trapped in the surface material. A microphone on board the lander will allow us to hear, for the first time in human history, the high-pitched hum of the winds on Mars.

The Psychologically Crucial Step: Sending Humans to Mars

For many, the sounds from Mars will resonate more deeply than any news about volatile compounds trapped in Martian soil. NASA had no plans to include a microphone on the *Mars Polar Lander* (since it had not sent any on previous missions such as those of the two *Viking* landers) until a private organization, The Planetary Society, mounted a successful campaign to add this small, inexpensive item to the mission. Founded two decades ago by Carl Sagan and Bruce Murray, The Planetary Society ranks among the largest of the advocacy organizations that promote space exploration. Most of this group's efforts have gone into urging additional funding for the robotic explorers that have unveiled the solar system to our eyes. Other groups, far more eager to add a human component to our efforts, have been far more outspoken in urging that we move as quickly as possible past automated spacecraft and begin serious plans to send humans to Mars.

Almost everyone agrees that human exploration of Mars should and will occur; the question is one of time scales and attitudes toward why we want to put humans on our neighboring planet. Automated explorers have considerable advantages over the human variety. Robots neither eat nor breathe; they have minimal requirements for energy consumption and for defense against the high-energy particles and radiation that bombard everything in the solar system not protected by shielding. With steady technological improvement, we can train robots to perform a host of tasks more efficiently than any human can. During the past three decades, robotic spacecraft have opened the myriad worlds within the solar system to human experience. We build them, we send them, and we can and do guide them once they arrive on a planet, a satellite, or an asteroid.

Enough already, cry those who dream of human explorers and Mars colonists. Take your liberal, robotic claptrap, your fearful, negative thoughts, retire to your diminishing room on Earth while we explorers leave the naysayers behind to soar into space, the hope for humanity. You represent the past, with its stuffiness and fear; we are the future.

This may be true. We can all accept the prediction that humans will eventually travel throughout the solar system, examining its objects as only humans on the spot can do, and will create colonies on some of the objects most useful to us while dreaming of journeys a million times farther to the planets that orbit other stars. The fact that some envision the first humans landing on Mars in about the year 2035 while others point to the year 2015 amounts to the most modest of disputes on the long map of history. A much deeper disagreement, however, deals with the attitudes that different people bring to bear on the objects within and beyond the solar system. From my perspective, the damage we have wrought on Earth cries out for us to exercise extreme caution before inducing extensive changes on other worlds. Far from being the common heritage of humanity, these objects belong to the cosmos. It is one thing to explore these worlds, leaving modest amounts of debris and causing some damage (both of which we might someday rectify) and quite another to plan wholesale colonization or restructuring of planets, moons, or even modest asteroids.

But we must provide rough justice to those who eagerly want to reach Mars as quickly as possible, not with spacecraft but in the all-too-human flesh. These are not hostile people; they want the best for humanity, including a chance to expand our physical, intellectual, and emotional horizons. Their leading figure for the past few years has been Robert Zubrin, a skilled engineer whose book *The Case for Mars* has galvanized those who want to send humans to Mars not eventually, not a generation from now, but in no more than a decade—the same length of time it took us to send humans to the moon once this goal was made a national priority.

But Zubrin and his associates in the Mars Society, which they created during the summer of 1998, do not view the United States or any other government as the key to reaching Mars. From their viewpoint, NASA remains an amazingly sluggish organization, wedded to the slow process of creating the International Space Station in Earth orbit, and then a base on the moon, before turning to the more exciting challenge of a human mission to Mars. The Mars Society aims to solicit and to obtain sufficient funds to send people to Mars without relying on

NASA to do the job. Some of its members are determinedly antigovernment and dream of living on a planet without laws and regulations, while others simply feel that governments move too slowly, and that their lives will expire without a chance to explore Mars.

Zubrin and his colleagues know perfectly well that at the present time the only voyagers going to Mars are automated spacecraft sent by NASA and ISAS, with an ESA mission likely a few years later. "We are probably going to have to start small, perhaps by raising enough money to put a hitchhiker payload on a NASA Mars mission, or [on] a European Mars mission; that might cost about ten million dollars," Zubrin says. "But if we could do that, I believe that we would have the international public profile required to raise perhaps $100 million to fly a full-up robotic Mars mission, comparable, say, to the *Pathfinder* mission. I believe at that point we would have the stature required to raise the billions to send the first people to Mars—either by ourselves or on a teamwork basis with NASA or other government-funded efforts."

In assessing potential public support for the Mars project, Zubrin brims with optimism, noting that when the *Pathfinder* spacecraft landed on Mars in the summer of 1997, NASA's web site providing updated information received "hits" from more people than the number who vote in the United States. "There are more people in this country and internationally that care about us going out there and pushing new worlds than care about all those other issues that appear to dominate the political landscape," he claims. "The people who are behind on this are the politicians." The crucial test of Zubrin's claim will of course arise when the Mars Society asks people to add their principal to the interest they have demonstrated in sending spacefarers to Mars.

The Mars Society has long-range goals that extend well beyond human exploration of Mars to its colonization. Zubrin writes, "Mars is an incredible object of scientific inquiry. But it is not just an object of scientific inquiry; it is a world. It is a world that has on it all the resources to support not just life, but someday a new branch of human civilization. . . . I think that what we have here is a chance to write a new chapter in the human story."

Promoting the space frontier as the key to the improvement and survival of human society, Zubrin recalls the importance of the American frontier, as emphasized by Frederick Jackson Turner and other historians (but not without criticism from still other historians, who claim that the frontier's effect on society has been exaggerated):

> What Mars is all about is the opportunity to establish a new branch of human civilization, in a new place, under new conditions, where people will be forced to innovate, and where they will also be free to innovate, because they will be free of the social constraints of the customs imposed from previously existing society, to a fair extent in any case. This is what frontiers have always offered. . . . Once people settle on Mars, there is going to be a tremendous push for development of more advanced transportation technologies. We will have nuclear propulsion, we will have electric propulsion, we will have fusion propulsion. We will have solar sails and magnetic sails, and these kinds of technologies will make travel to Mars routine. . . . Perhaps these sorts of propulsion systems will give us at least a marginal capability for reaching for the stars.

Like many others who dream of colonizing new worlds, Zubrin rates the psychological effects of this effort even higher than the technological breakthroughs that colonization offers. "In the long term," he says, "I think it is absolutely imperative that we move out to the planets and to the stars. I really don't think that human civilization can remain viable if it remains within a domain in which it has settled all the major environmental challenges of that domain. That is a formula for stagnation. If we are to remain progressive, if we are to remain flexible, if we are to continue to grow, then we must do exactly that, we must grow and spread."

A more balanced view of space colonization comes from Freeman Dyson, a physicist at the Institute for Advanced Study in Princeton, New Jersey, and one of the great scientific thinkers and imaginers of our time. Dyson draws an obvious distinction when he states that

> the main lesson that I draw from the history of space activities in this century is that we must clearly separate short-term from long-term aims. The dream of expanding the domain of life from Earth into the universe makes sense as a long-term goal but not as a short-term goal. The practical feasibility of cheap human voyages and settlement of the solar system depends on fundamental advances in biology. Any affordable program of manned exploration must be centered on biology, and will have a time scale tied to the time scale of biotechnology. A time scale of fifty years is probably reasonable. This is roughly the time it will take us to learn how to grow warm-blooded plants [which can produce heat and regulate their temperatures, as a few plants do on Earth]. . . . They are only the first of the thousands of diverse new species

that will be required to create viable ecologies in the places where humans may wish to go. A biological technology mature enough to create warm-blooded plants will also be able to take care of other ecological problems, on Mars or on Europa or even at home on the Earth.

The Terraforming of Mars

When Zubrin and the Mars Society look a century or more into the future, they see humans on Mars, not as explorers or colonists but as inhabitants, many of whom will have been born on our neighboring planet. To support a significant human population, Mars must undergo a serious reconfiguration, one that will cover its surface with vegetation and add oxygen to its atmosphere. This restructuring goes by the name "terraforming," a word first used in science fiction but now adopted by the space exploration community as a reminder that these proposed changes would make Mars resemble Earth, also known as Terra.

Of all other planets, Mars is most ripe for terraforming because it resembles Earth most closely, possessing a solid surface, a significant atmosphere, and Earth-like temperatures. Venus, much closer to Earth in size, has a thick carbon dioxide atmosphere that traps heat from the sun so efficiently that its surface bakes at 700 degrees Fahrenheit even on the planet's nighttime side. Mercury, with no atmosphere at all, has a surface almost equally hot because it orbits so close to the sun. The giant planets Jupiter, Saturn, Uranus, and Neptune have layers of gas, thousands of miles deep, atop their solid cores, offering no good prospects for life on surfaces or in oceans. Pluto, at forty times Earth's distance from the sun and far smaller than our moon, amounts to a giant comet, forever orbiting in a celestial deep freeze hundreds of degrees below zero. Mars, in contrast, has temperatures that rise to almost 60 degrees Fahrenheit in high summer and fall to "only" about -150 degrees Fahrenheit during Martian winter nights.

If we could somehow thicken the Martian atmosphere significantly, we could produce two key changes on Mars. First, a thick atmosphere would trap more solar heat, raising the average surface temperature by perhaps ten degrees Fahrenheit in the daytime and more than that at night, as the thicker atmosphere would provide thermal insulation after sunset. Second, the increased atmospheric pressure would allow liquid water to exist on the surface of Mars. Since a higher average temperature would vaporize some of the water now frozen in the polar caps, a thick Martian atmosphere would eventually produce a complete water cycle with rainfall, runoff into streams and ponds, evaporation, and further rainfall. Even so, water could never become anywhere nearly as abundant on Mars as it is on our planet; indigenous Martians might therefore create a complex series of canals and oases, transforming Lowell's vision of Mars from misconception into precognition.

The key to terraforming Mars would be the introduction of plants, genetically engineered from those that grow in the colder regions of Earth, that can survive the harsh Martian conditions. Since plant life requires carbon dioxide and the Martian atmosphere, thin though it may be, consists mainly of carbon dioxide, one obstacle to this plan can be regarded as already largely overcome. A plant with sufficiently long roots for tapping small amounts of water in Martian soil could flourish despite the permafrost conditions, perhaps by evolving a tiny greenhouse of its own to retain heat during the cold Martian nights. As the atmosphere becomes thicker, these plants could take advantage of the increased amounts of water in the atmosphere and the emergence of liquid water on the surface.

Since plant metabolism produces oxygen, Martian plants in sufficiently large numbers could eventually make the atmosphere relatively oxygen-rich. This atmospheric oxygen would allow animals to breathe, perhaps as if they occupied high-altitude ecological niches on Earth, and would also provide shielding against ultraviolet radiation from the sun. Like Earth's atmosphere, which protects terrestrial forms of life that have evolved beneath it, Mars's barrier to ultraviolet radiation would consist of oxygen atoms, oxygen molecules (pairs of oxygen atoms), and ozone molecules (triplets of oxygen atoms, each made when an oxygen molecule combines with an oxygen atom).

Terraforming would therefore bring heating and greening to Mars. A sufficiently large initial impulse of heat, produced by artificial means, could release significant amounts of carbon dioxide and water vapor from the polar caps, thickening the atmosphere and increasing the Martian "greenhouse effect," which in turn would increase the temperature on Mars still further. The greening of Mars by fields of vegetation would enrich the Martian atmosphere in oxygen and allow it to form a high-altitude shield against solar ultraviolet radiation. Over a time span estimated to be from a few centuries to many millennia, the red planet might be transformed into "that other Eden, demi-paradise" that Shakespeare called Mars in another guise. Since the Martian surface has an area equal to Earth's total land area, this would represent no small addition to the human field of activity—a lebensraum whose acquisition may seem to be humanity's manifest destiny.

Some who read of terraforming will object to the concept of remaking a planet in Earth's image. In rebuttal, terraforming enthusiasts deride conditions on Mars today. "[Terraforming] evokes images of colonialist arrogance, a perversion of the state of a planet into something utterly alien to its nature," wrote John Lewis, an expert on planetary science, in his book *Mining the Sky*. "But what we envision for Mars is an interference more akin to refurbishing and rewinding a fine old clock we have found that ran down many years ago. We are restoring Mars to where it was when it slipped off the rails of planetary evolution. Rather than terraforming, we are 'areo-reforming' Mars—restoring Ares, the Greek god of war, to his rightful dominion. We cannot, and arguably should not, attempt to make Mars into another Earth. But we can give it new life, life that will prosper and proliferate in proportion to how well it is adapted to the restored, reinvigorated Mars."

Lewis here propounds a very subtle distinction. If we decide that Mars has "slipped off the rails," and that we should "restore" it, have we not engaged in planetary engineering? If a planet has changed from state A to state B and we then change it back to state A, are we not as colonialistically arrogant as if we decided to install state C? These are issues worth thinking about; they may soon—if Zubrin, Lewis, and others who are eager to be out and doing in the solar system have their way—become a hot political topic. As we examine the pros and cons of reworking Mars, we can also apply a similar analysis to human activities that may soon occur on those little known but potentially important objects, the asteroids.

Chapter 6

Mining the Asteroids

The host of objects in orbit around the sun include its well-known set of nine planets, plus more than fifty satellites of these planets, much larger numbers of small objects called asteroids, and still greater numbers of comets. Asteroids, whose name means "starlike," appear pointlike in a telescope, just as stars do, but in fact are cold, primordial pieces of the solar system that never became part of a planet during the era when larger celestial objects assembled themselves. Most of the smaller moons in the solar system were once asteroids, eventually captured by a planet's gravity into subservient orbits around the planet rather than maintaining individual orbits around the sun. Comets, even older than asteroids, are lumps of ice and carbon dioxide frozen around dust, grit, and rocks, and are famous in astronomical lore because when they make a relatively close approach to the sun, solar heating releases some of the frozen material which streams in the direction away from the sun to produce the comet's spectacular, gauzy tail.

Which Objects Come Close to Earth?

Comets and asteroids, including those that may come close to or even strike Earth, follow a general rule: Larger objects are rarer than smaller ones. As a result, close encounters with large asteroids or comets occur far less frequently than with small ones. This rule extends all the way down to celestial objects (called meteoroids) less than a few yards across that continually strike Earth in great numbers.

Fortunately, our atmosphere provides a barrier that meteoroids penetrate only at the cost of the extreme heating they undergo as they plow through the atmospheric gas. This heating destroys the smallest meteoroids, which flash in our skies as "meteors" by the millions every night. A typical meteor or "shooting star" is a pinhead-sized particle of primordial solar system material heated to the point that it glows, dozens of miles above Earth's surface, with a momentary brightness visible far below. Larger meteoroids, at least the size of a breadbox, can partially survive this atmospheric heating to be found by curious humans who label them "meteorites." Most meteorites, hard to distinguish as extraterrestrial objects, consist of rocky material, but a small number of them, no more than 2 or 3 percent, are mostly iron and other metals. This proportion seems to apply to asteroids in general: Most of them are hunks of rock, but a few are mainly metal. The words "meteoroid," "meteor," and "meteorite" all derive from the Greek word for "wind," a reminder that the extraterrestrial origin of meteorites became evident only during the first part of the nineteenth century. On the other hand, the Greek word *sideros*, which became the Latin word *siderus* ("star"), also provided a word for iron.

These derivations suggest that ancient civilizations may have first encountered iron in the form of meteorites and assigned it a heavenly origin.

Like their much smaller cousins, meteoroids, some asteroids come close to Earth. Although the great majority of the solar system's asteroids orbit the sun between the paths of Mars and Jupiter and therefore always maintain distances from the sun that are between two and four times the Earth-sun distance, a small minority, which nevertheless includes thousands of asteroids of considerable size, orbit at roughly Earth's distance from the sun.

Astronomers group these "near-Earth asteroids" into three categories, calling them Apollo, Atens, and Amor asteroids after the most prominent exemplars of each class. Apollo asteroids all move along elongated orbits that give them an average distance from the sun greater than the Earth-sun distance, which averages 92.9 million miles and varies from this average by only plus or minus 1.7 percent. However, Apollo orbits are all sufficiently elongated that their closest approaches to the sun bring them inside Earth's orbit. This makes Apollos "Earth-crossing asteroids" capable of some day colliding with Earth. In contrast to Apollos, Aten asteroids all have average distances from the sun that are less than Earth's and can cross Earth's orbit only because they likewise move in significantly elongated orbits, which at their maximum distance from the sun carry them farther from the sun than our 92.9 million miles. Amors are not Earth-crossing asteroids but only near-Earth asteroids whose elongated orbits take them from the heart of the asteroid belt, about three times farther than Earth from the sun, inward to much lesser distances just 10 to 40 percent greater than the Earth-sun distance. Hence we may someday meet directly with an Apollo or an Aten, but to rendezvous with an Amor, we would have to make a journey of our own.

The Danger of Impacts

Because the orbits of Apollo and Aten asteroids intersect Earth's orbit, these objects sooner or later may occupy a time and place that coincides with Earth's and collide with our planet. We know that such impacts have occurred in the past, and so it behooves us to worry about future collisions as well. In categorizing these worries, we must look beyond Apollo and Aten asteroids, of which we have detected a few hundred while another few thousand are estimated to await our discovery. Because none of these objects has a diameter greater than one or two miles, none of them poses a global threat to Earth. The greatest impacts have occurred, and will occur in the future, when asteroids or comets with diameters of ten miles or more strike Earth. So far as asteroids go, these rogue impactors are probably members of the asteroid belt, circling the sun at two to four times Earth's distance, that are diverted into new, highly elongated orbits by close encounters with other asteroids. If we wait for a sufficiently long time, one or more of these encounters will involve an Earth-crossing asteroid large enough to destroy most or all of life on Earth if it should collide with our planet.

Meteoroids the size of a house collide with Earth on human time scales. The remnant of one such collision, the milewide Meteor Crater in Arizona, was created at about the time that humans first reached North America and testifies to the destructive power of even a rather

An antimatter-powered spacecraft sets off on an interstellar mission. Such technologies may one day allow us to reach 10 percent the speed of light, yet we would still require a human lifetime or more to reach the nearest star to the sun. *(M2 Art)*

Overleaf: Setting off from an orbiting space station, a mission to another star system would certainly ignite a global debate over who we are as a civilization and how unique our world will turn out to be. *(M2 Art)*

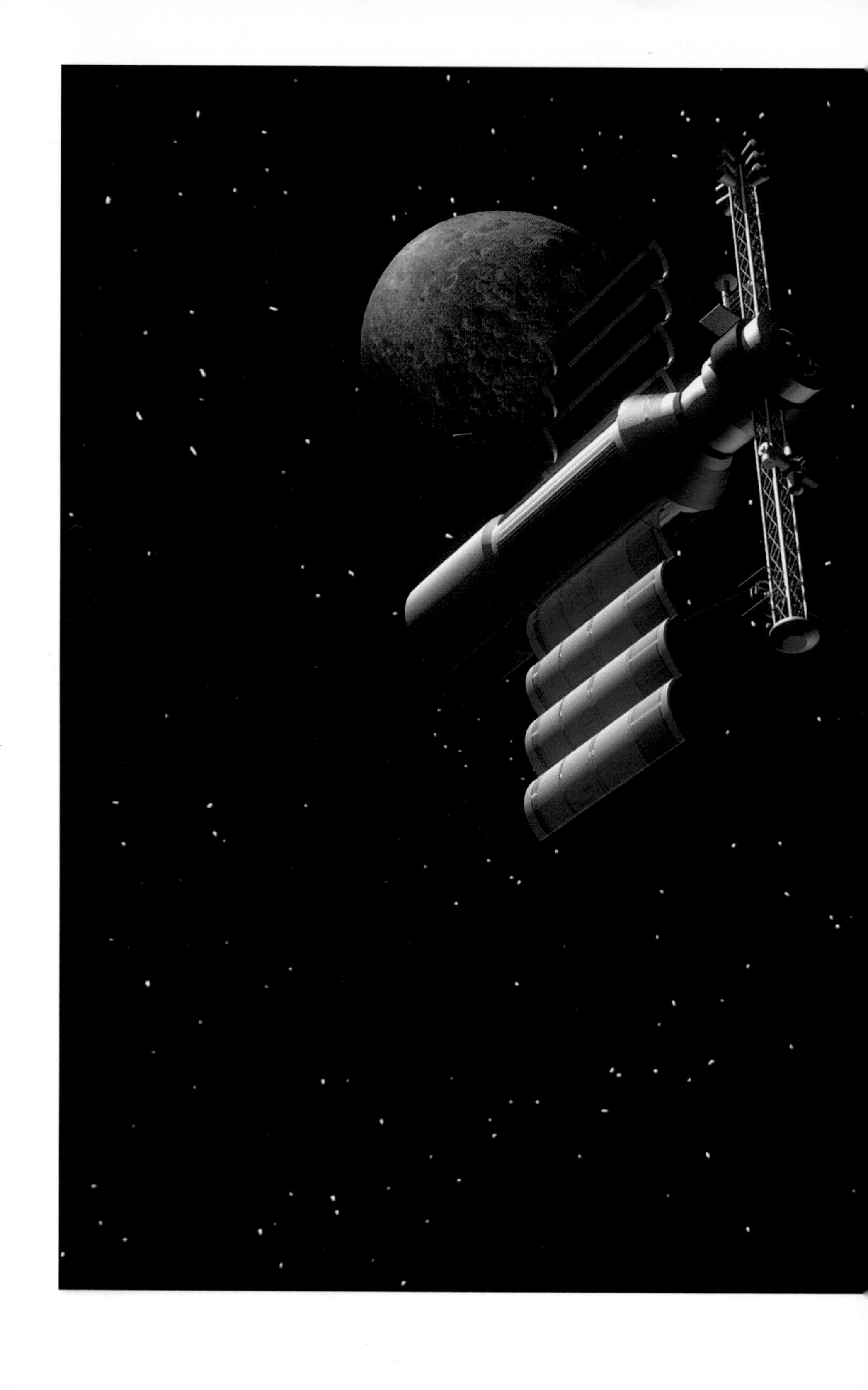

NASA currently has invested much of its resources in the International Space Station. In addition, NASA is working on a variety of advanced propulsion projects that promise to radically expand our reach into space.

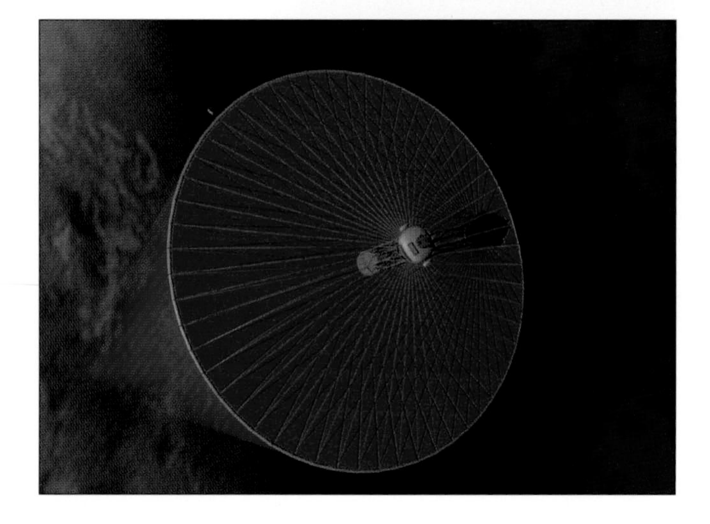

A "lightcraft," for example, will soar on microwave energy beamed from an orbiting solar power station. The craft will use the energy to propel itself upward, and to create an airspike that literally melts away air resistance. *(Senter Reinhardt)*

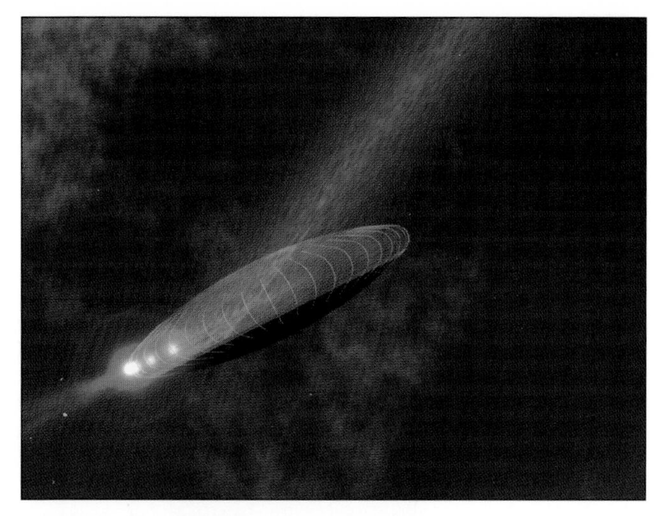

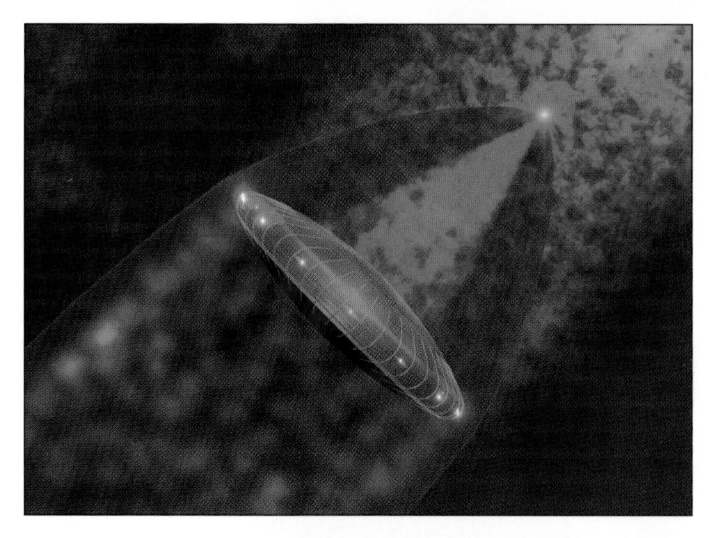

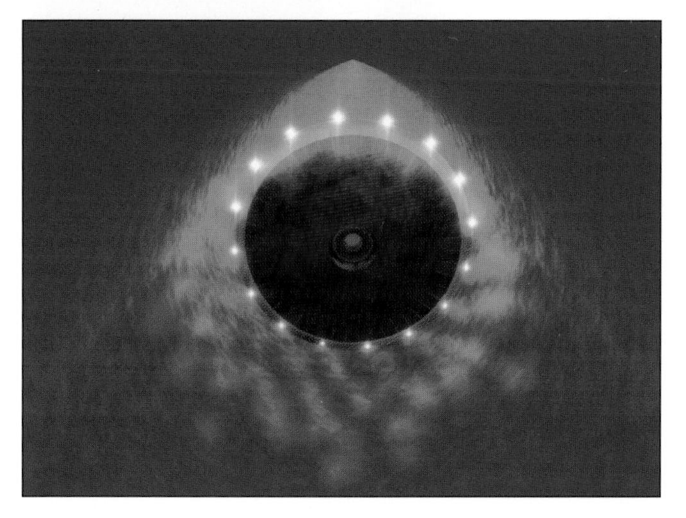

Solar Thermal Upper Stage, for shuttling around in Earth orbit.

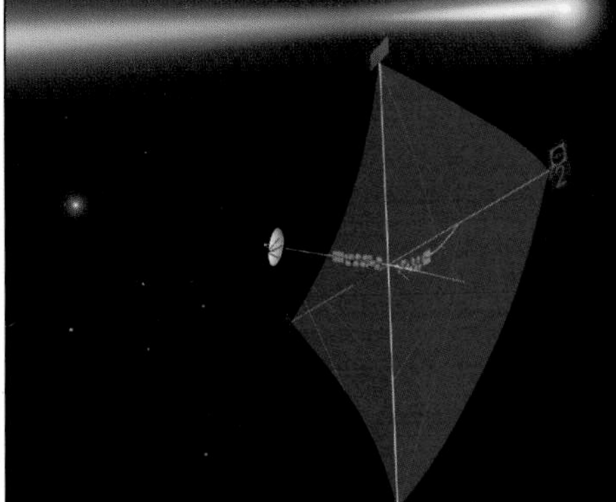

A solar sailcraft, pushed through space by the pressure of sunlight.

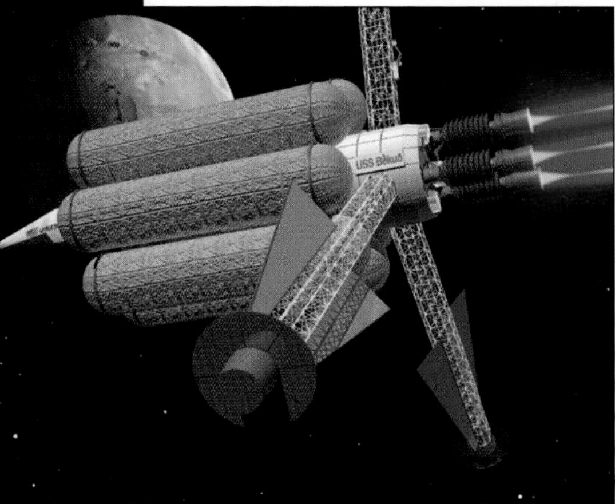

Superheated plasma is one of the most efficient and volatile fuels currently being tested.

Fusion vehicle concepts for the far future promise safe and highly efficient spacecraft.

Long-distance propulsion systems now on the drawing boards may make interplanetary travel as routine as flying across the Pacific Ocean. *(Marshall Spaceflight Center)*

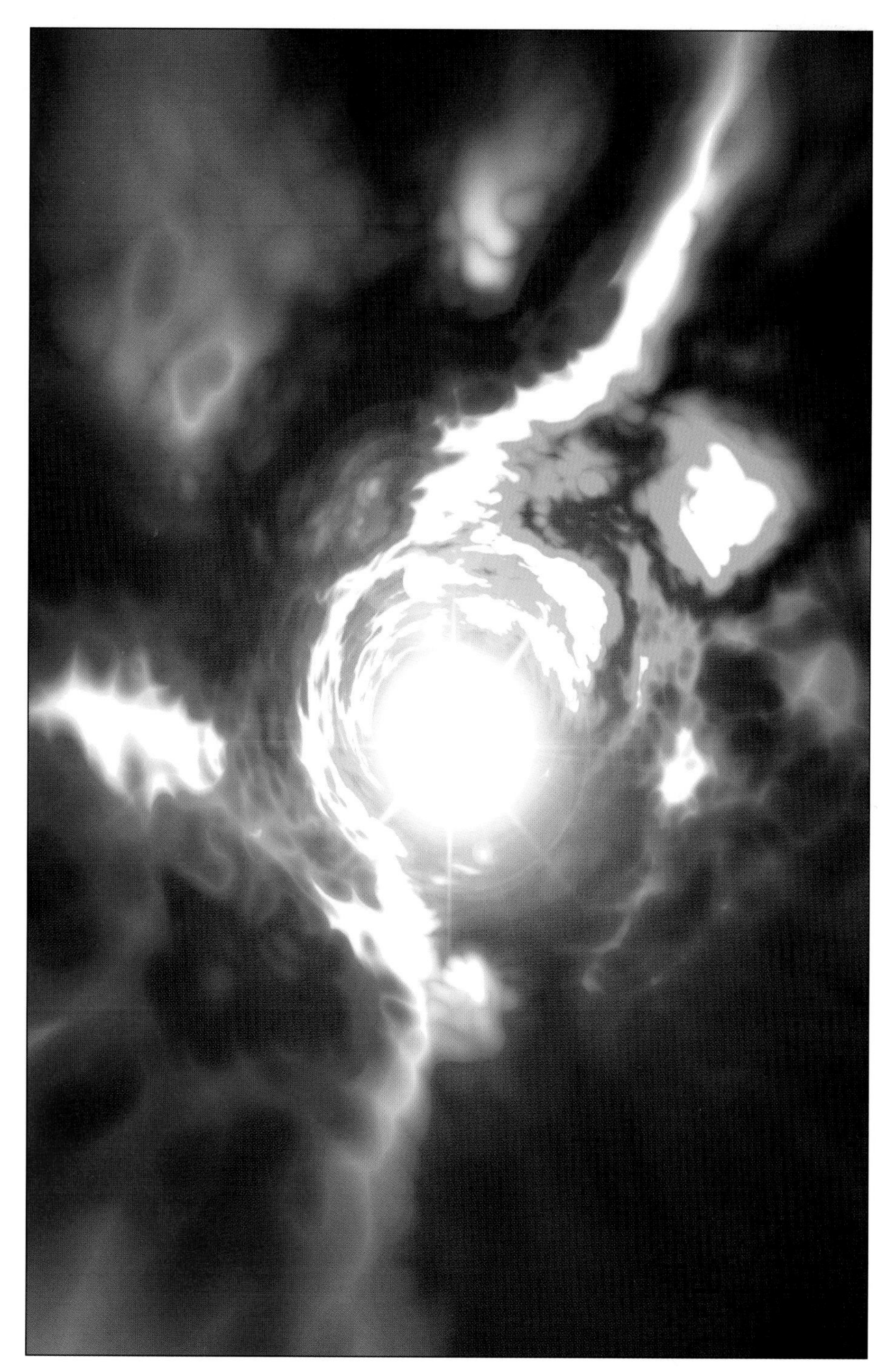

Starlight pours into the mouth of a wormhole, the ultimate passageway in space. Wormholes are theoretical places where space folds back onto itself, offering rapid but highly dangerous shortcuts to other times and places. *(M2 Art)*

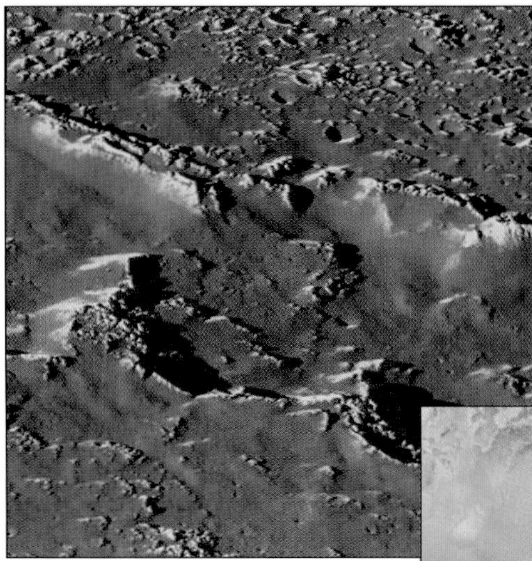

Callisto, one of the most heavily cratered objects in the solar system, may have a salty ocean beneath a surface layer of compressed rock and ice.

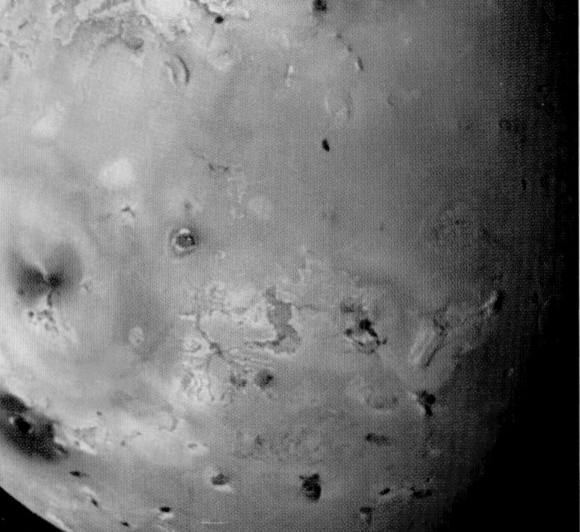

The changing surface of Io shows recoating on the most volcanically active object in the solar system. Internal heat and pressure produce eruptions over 300 kilometers high.

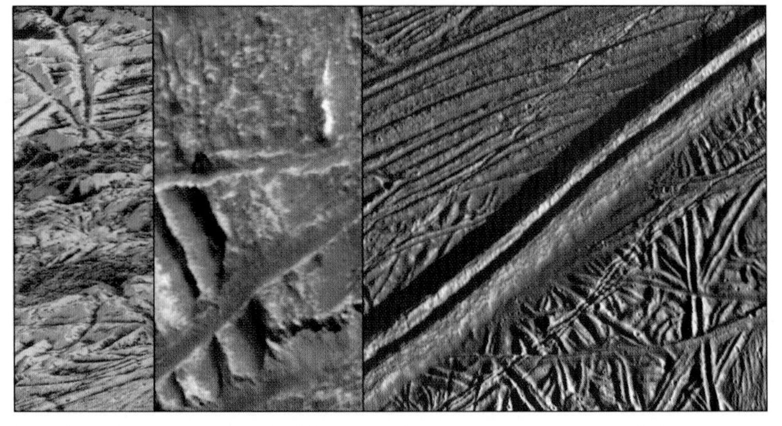

Surface fractures and ridges on Europa appear on this moon's icy crust. Beneath the ice, a warm subsurface ocean may harbor life.

The moons of Jupiter are potential destinations for explorers of the future, on scientific missions to discover primitive life forms or to understand the origin of our solar system. *(Jet Propulsion Laboratory)*

Jupiter, seen from the surface of Europa, shows the challenge of human space exploration: extreme temperatures, deadly radiation, and possible equipment failure far from Earth. *(M2 Art)*

A lunar colony could offer astronomers a clear view into deep space or the opportunity to mine rare minerals. The colony might provide the flagship for a far-flung space infrastructure. *(M2 Art)*

An asteroid-like colony, fitted with its own propulsion system, could travel for many generations until it eventually found another solar system. *(M2 Art)*

A concept for a Lunar Hotel, envisioned by the Hilton International hotel chain. *(Inston Design International)*

The Martian landscape of the future? New challenges will surely produce new opportunities as the rule of law and possibly private property rights are established in space. *(Jet Propulsion Laboratory, Graphic by Gary Looft)*

An orbiting space hotel, proposed by the Shimizu Corporation. The hotel rotates to produce artificial gravity, allowing a good night's sleep before the pleasures of the zero-g gymnasium. *(Shimizu Corporation, Space Systems Division)*

A moon crater could provide a fine playing field for an old-fashioned game of baseball. As humanity spreads into space, what earthly pastimes will we bring along? *(Rhinoceros Visual Effects)*

Mars as it is today—dry, cold, and apparently lifeless.

Mars as it might appear after terraforming. The introduction of greenhouse gases could turn the red planet green, creating lush new territory for planetary pioneers. *(Jet Propulsion Laboratory)*

modest object striking our planet at several miles per second. Larger objects moving at similar speeds will of course create greater destruction. As long as the impacting object has a diameter no greater than one or two miles—the size of the largest Apollo or Aten asteroid—this damage, though immense, will remain a basically localized phenomenon. Such impacts presumably have occurred throughout Earth's history at intervals of a few million years. No one would want to be close to the site of one of these impacts, but Earth as a whole will survive relatively unscathed.

As the size of an impacting asteroid or comet surpasses a few miles, however, we enter the realm of Earth-changing events. These collisions occur in the same random manner as impacts with smaller objects, but with longer time scales because Earth-crossing objects of these sizes are scarcer than smaller ones. The geological record shows that at random intervals of about 20 to 40 million years, our planet has been impacted by an object sufficiently large to alter the course of life on Earth. These great collisions have provoked "mass extinctions," the wholesale dying of a large portion of all living species, apparently as a result of debris being thrown above the atmosphere, which produces a worldwide darkening that lasts for months. Only the largest impacts create such a large hole in the atmosphere and eject so much grit through it that Earth's atmosphere receives a coat of fine dust that takes months to settle back to the surface and blocks most of the sun's light until it does so.

The most famous of these mass extinctions (but not the greatest one in the fossil record—that distinction belongs to the "end-Permian extinction" about 250 million years ago) occurred 65 million years ago when an asteroid judged to be about six miles wide struck Earth at the northern end of the Yucatan peninsula, killing all the dinosaurs and a great number of other types of creatures then flourishing on Earth. Since that great era of dying, the fossil and geological records speak of only one more collision with an asteroid at least a few miles wide, about 30 million years ago.

If we reasonably extrapolate what has happened on Earth in the past into our planet's future, we can calculate that on average, asteroid impacts will cause several hundred deaths per year. In other words, every 30 million years or so, we all die. Large asteroids—those more than five miles wide—do almost all the damage. Smaller asteroids exist in greater numbers than larger ones and therefore strike Earth more often, but they produce only localized disasters and with any luck fall into the seas, raising only a few enormous waves. The objects capable of extinguishing most life on Earth are those with diameters hundreds of times larger than the impactor that produced Meteor Crater in Arizona.

A wise United States Congress has considered funding efforts to catalog the larger near-Earth asteroids, the Atens, Apollos, and Amors. And groups of concerned scientists have considered what to do if we should detect one of these that actually seems destined to collide with Earth during the next few decades. These possibilities have furnished a fundamentally sound basis for action movies, though the remedy for a predicted impact will require far greater efforts than the films offer. For example, in the movie *Armageddon,* our leaders meet the threat of impact by a Texas-sized asteroid (which has strangely escaped detection by astronomers) by sending a team of roughnecks to drill half a mile into it and place a nuclear explosive to blow the asteroid apart. Why drilling for only half a mile should do the trick, and why pieces of the asteroid would not themselves hit Earth, remain poorly answered questions. In real life, what worries scientists most about dealing with an impacting asteroid in this manner is the possibility that we might accidentally change a near miss into a collision or divert a large portion of an asteroid onto our planet.

Advances in technology will improve our ability to detect potential impactors with Earth and to deal with the threat from near-Earth objects. At the very least, obtaining a complete catalog of all the larger objects that might strike Earth will help us decide how to formulate long-range plans to deal with the asteroid problem. This knowledge will also prove highly useful for future efforts to send astronauts to some of the closer asteroids, not to blow them apart but to dig them up.

Asteroids as Sources of Raw Materials

In terms of mass, the solar system's millions of asteroids, billions of comets, and trillions of meteoroids amount to almost nothing. The planet Jupiter probably contains more mass than all these objects combined, and even the largest of them—the asteroid Ceres, which is as large as Texas—has less than one thirtieth of the mass of our moon. Yet these asteroids may offer spacefaring entrepreneurs their greatest possibilities for acquiring wealth.

How can objects as small as asteroids offer greater opportunities than the moon or Mars does? The answer lies in two properties possessed by asteroids, both directly related to their relatively modest dimensions.

First, because asteroids are small they exert only modest gravitational forces. This makes it far easier, in energy terms, to land on and to take off from an asteroid than to do so on or from a large moon or a planet. On the balance sheet of finding and mining mineral deposits elsewhere in the solar system, the low gravity of asteroids argues heavily in their favor because any mining operation will require repeated trips to and from the object being exploited.

Consider an asteroid a mile across, the size of a modest town's center, such as the student core of Berkeley, California, or Princeton, New Jersey. This object consists of rock and metal, with a composition that in most instances resembles that of Earth's crust. The mass in the asteroid approximates ten billion tons, most of which consists of rocks—familiar compounds of silicon, oxygen, and other elements like those that form the rocks near Earth's surface. All asteroids, however, contain a small fraction of metals, and a small percentage of them, as we have noted, consist mainly of metal.

Let us assume that our asteroid ranks in the latter category. In that case, it should include billions of tons of useful metallic elements, mainly aluminum, iron, nickel, copper, zinc, and cobalt. A small proportion of the total should consist of more valuable elements such as molybdenum and lead, and a still smaller fraction (but one that still amounts to thousands of tons) should be composed of even more valuable elements, including platinum, iridium, palladium, mercury, silver, gold, and uranium. Thus the asteroid represents an "ore body" richer than anything ever discovered on Earth, such as the famous deposit of metals (itself perhaps the remnants of an asteroid that struck Earth hundreds of millions of years ago) near Timmins, Ontario. If iron sells for $100 per ton, the asteroid contains trillions of dollars worth of iron alone. If platinum sells for $100 million per ton, which it does, then the asteroid's platinum, amounting to hundreds if not thousands of tons, offers trillions of dollars more to those who can mine it, refine it, and bring it to Earth.

This operation certainly sounds like a promising one. Those who dream of exploiting the mineral wealth in orbit around the sun find it appropriate that we should begin with the Earth-crossing asteroids, the ones that might someday strike Earth. How marvelous it would be if, long before any such damaging impact could occur, we could strip the danger away, turning the asteroids' innards to our own uses. Asteroid miners would not only offer no apologies for changing our near-Earth environment but would indeed merit high praise for preventing destruction on Earth while producing a net economic advantage.

In assessing the payoffs from an expedition to mine an asteroid, we must analyze, as we would for any mining operation, the crucial issues of how the cost of extracting the sought-after material compares to the economic reward of doing so. Simply citing the total value of the extracted material hardly settles the issue. After all, Earth's core, thousands of miles below the surface, contains far more iron, nickel, copper, and cobalt (not to mention silver, gold, platinum, palladium, and uranium) than all the asteroids combined, yet we recognize instinctively that the costs of mining the core outweigh the economic benefits. Similarly, the sun contains far more of the elements that we seek than all the planets combined, but the obvious difficulties of extracting this material make it evident that we need not debate mining the sun.

What, then, of mining the asteroids? The costs of sending a mission to investigate a solar system object depend almost entirely on two factors: the distance to the object, and the gravitational forces that must be dealt with to make a trip to and from it. Of these two factors, the distance is of lesser importance, the more so if the mission is largely or wholly automated so that the costs of maintaining a crew on a lengthy voyage can be minimized. What counts most is how much energy must be invested on launch, how much on reaching the object of the mission, how much on leaving the object, and how much on returning to home base. The greater the gravitational force exerted by either the site of the launch or the object of the mission, the greater the energy costs. These costs translate directly into real monetary expenditures because—at least for the foreseeable future—the energy must come from rockets that burn chemical fuel, all of which must be carried by the rocket on its journey out and its journey home.

These energy facts lead to important conclusions. First, a mission to mine an asteroid can proceed far more cheaply if it begins at a lunar colony than if it starts from Earth's immediate vicinity. In the former case, the mission must overcome a gravitational force only one sixth as strong as Earth's. The gravitational force from the asteroid itself is so far below that of the moon as to be almost negligible: A person on the surface of a two-mile-wide asteroid experiences a gravitational pull only about one-thousandth of that on the moon's surface. Thus a one hundred-pound astronaut would weigh only 16 pounds on the moon and an ounce and a half on the asteroid, where a properly slanted ramp could serve as a launch pad into space for an explorer equipped with a bicycle.

The gravitational force on the surface of a one-mile-wide asteroid would be just right to keep newly extracted material from flying into space as it is mined, while allowing mining machines to operate more easily than they do on the surface of a much larger object, where gravity represents an obstacle in bringing material to the surface. Of course, the chemical bonds that hold rocks and metal together in an asteroid have the same strength on large and small solar system objects, and so the actual cutting required to extract minerals would require roughly the same expenditure of energy.

Our second conclusion likewise deals with the impact of gravitational forces. The cost of using the material extracted from an asteroid is greatest if we seek to return it to Earth, less if we want it on the moon, and least of all if we plan to use it in space, either to build habitats in space or for similar construction on the low-gravity surface of an asteroid. Because it costs very little to move the material in these cases, almost the entire cost lies in creating and operating the asteroid excavation. As a result, asteroid mining will truly come into its own once it supplies not Earth or the moon but rather operations in much lower-gravity environments. Asteroidal mineral extraction works best by refining one asteroid to support activities on that object or on another asteroid that periodically passes close by.

Picture, then, a mining camp on a mile-wide, mineral-rich asteroid. Established during one of the asteroid's close passages by Earth, the camp might shelter a dozen hardworking miners, maintaining the operation of giant rock cutters that strip away layer after layer of the asteroid. Other machines would load the freshly cut ore into small rockets for a trip to the moon, for in this scenario, the asteroid's innards furnish raw material for lunar factories. The extent to which separation of the ore into its different components occurs on the asteroid itself, or in a larger facility on the lunar surface, would depend on just which components are involved and how large an operation is to be maintained on the asteroid. But since the asteroid's orbit carries it far from Earth—on average, the miners would be as far from Earth as the sun—while the lunar base remains close to us, it seems likely that a far larger population could be more easily maintained on the moon than on the asteroid. This in turn suggests that unless we can build highly automated plants on the asteroid itself to separate the different metals, most of the ore might be sent to the moon for refining. In either case, our goal (conceptually at least) is the creation of neat piles of refined elements, ranging from the rarest and most valuable (such as gold and uranium) to the most abundant though still valuable (such as aluminum and iron). The largest pile of all, to be sure, will be the tailings from the mine, which will accumulate both on the asteroid and at the site of separation.

When Does the Mining Begin?

How far are we now from seizing the day and beginning to extract the valuable minerals that asteroids offer to us as they orbit past Earth? The answer, as always, depends on a number of issues, including the value of the minerals and who will do the seizing.

If we mark the beginning of asteroid mining by the start of construction on a spacecraft designed to achieve this goal, the era of extraterrestrial mineral exploitation has begun. The year 1999 has seen the initial fabrication of the *Near Earth Asteroid Prospector (NEAP)* destined for launch in 2001, the first year of the new millennium (for calendar conservatives). *NEAP* represents a bold break with past attitudes toward exploring the solar system, which featured government-funded research without any visible economic payoff. In contrast, *NEAP* operates without governmental support as the creation of SpaceDev, a corporation founded with the goal of profiting from space exploration.

SpaceDev's immediate target is Nereus, a milewide near-Earth asteroid whose orbit will bring it into a favorable position in the spring of 2002, to be reached by a spacecraft launched from Earth about a year earlier. Like most asteroids, and unlike the metal-rich ones that offer wealth beyond the dreams of avarice, Nereus consists mainly of rock and ice. Such asteroids do offer the possibility of extracting oxygen, ice, and other compounds highly useful to future space travelers, if not as valuable on Earth as the contents of the rare metallic asteroids. SpaceDev's plans to send a spacecraft to Nereus aim at testing the concept rather than achieving an economic return; the corporation also plans to add to its income by selling space for scientific instruments to any government or private group that wants to pay for it.

The driving force behind SpaceDev is a middle-aged entrepreneur named Jim Benson. A decade ago, Benson grew rich from a company called Compusearch, which made one of the first Internet search engines. Like Robert Zubrin with his vision of planting humans on Mars, Benson has a firm conviction that the space frontier offers the best hope for humanity to realize its full potential. "A space economy is the only thing that is going to get humanity into space," he asserts, because taxpayer subsidies will never bring us more than shuttle flights. "If people want to go to space to stay, space has to pay. So we have to find things within space that make money, and allow people to live and work in space. . . . If we can use our probes as transportation, we have three key ingredients of space industrialization: energy, communications, and transportation. That is the infrastructure that is required for commercializing space."

Benson has no background in astronomy, but his advisers include some of the world's experts on comets and asteroids, as well as former NASA officials who became convinced that the future of space exploration lies with nongovernmental projects. NASA's administrator, Dan Goldin, has high admiration for Benson's goals, which resonate with the "faster, cheaper, better" mantra he has introduced at the space agency. As far as the voyage to Nereus goes, this makes perfect sense, since even Benson regards this first SpaceDev flight as strictly a scientific mission that will drop instruments on the surface of Nereus and relay data to Earth from these instruments.

A successful launch to Nereus by SpaceDev will lead to an intriguing result: an exploration competition between private entrepreneurs and governmental space agencies. The latter include NASA and ISAS, which have agreed to use a Japanese launch vehicle to send a spacecraft to Nereus late in 2002. Unlike SpaceDev's effort, the NASA-ISAS mission will achieve an actual landing on the asteroid in September 2003. Then a miniature, two-pound rover, the NASA contribution to the mission, will move over Nereus's surface, collecting samples that will be returned to Earth in January 2006.

All these projections remain, of course, subject to the vagaries of human engineering and cosmic luck. The first mission dedicated to examining an asteroid, involving NASA's *NEAR* mission, was launched in 1996 on a long, looping trajectory that carried it close by the asteroid Eros in December 1998. Eros, one of the largest near-Earth asteroids and the most famous asteroid with an Earth-crossing trajectory, is a battered, potato-shaped hunk of rock about twenty-five miles long and nine miles across. (Asteroid experts like the phrase "potato-shaped" because it describes almost every nonspherical object. Only the largest asteroids, with diameters of a few hundred miles or more, have gravitational forces sufficiently strong to render the object spherical, like the planets and their large moons.)

NEAR was designed to match orbits with Eros for a year, approaching its surface within ten to twenty miles for close-up studies as the asteroid and spacecraft orbited the sun. Unfortunately, in late 1998 *NEAR* underwent an aborted "rendezvous burn" as it started its engines to match its orbit with the asteroid's. As a result, *NEAR*'s first quick flyby brought it no closer to Eros than twenty-five hundred miles. The complications of maneuvering in space with limited supplies of fuel will delay *NEAR*'s "true" rendezvous by more than a year, until May 2000, when NASA scientists hope to achieve the actual matching of orbits, allowing the detailed observations of Eros for which *NEAR* was designed.

Who Owns the Universe?

Thus the first private spacecraft mission to an asteroid will be followed by the first international asteroid mission, launched by cooperating governmental agencies. NASA and ISAS envision only scientific studies, while SpaceDev seeks to exploit the mineral wealth of asteroids. Does this competition (which extends even to acronyms, as SpaceDev's *NEAP* rivals NASA's *NEAR*) presage a struggle over who will have the opportunity to explore and to profit from the celestial objects around us?

For Jim Benson, the answer is clear: "Once we have completed the science [with SpaceDev's flight to Nereus], then we are going to have a little fun. We are going to say that this was a private company and it was privately financed and we landed on that little planetary body and we are going to claim that we own that body. I think it is extremely important to create a precedent for private property rights in space."

For those who believe that the key precedent lies in the United Nations' Outer Space Treaty of 1967, which declared celestial objects the common heritage of mankind, Benson has a ready answer. "The Moon Treaty [of 1979]... tries to reserve all of resources in space for all of humanity, making it impossible to earn a profit, because if you did, you would have to distribute it to all six billion people in the world. That is why it was never ratified. The [Outer Space Treaty] has been ratified by all the major spacefaring nations, and it does not allow countries to claim sovereignty over any planetary body, but it says nothing about property rights, or private ownership in space." As Benson correctly observes, the broad principles embodied in the Outer Space Treaty failed to receive implementation in 1979, when the major powers refused to ratify the Moon Treaty. As he also notes, this leaves the status of activities on other celestial objects in considerable doubt, crying out for a ruling from a supreme court—if only we had one for activities beyond Earth.

Were Benson on that court, the ruling would not be in doubt, for he insists that celestial objects in orbit around the sun (and, by an easy extension, throughout the universe) belong not to governments but to individuals or corporations, presumably those who first land on them to claim their rights. "I believe that because no internationally ratified treaty even mentions property rights in space, someone simply has to set a precedent and then let public opinion and the world's legal bodies arrive at a solution. I think we have, we will have standing because we are the ones in space. And I believe that those who are in space are the ones who ought to be setting the precedents and making the laws that will regulate what happens in space, not some theoretical body of the U.N., which, you know, has really in my mind no standing whatsoever."

In their individualistic approach to the exploitation of space, Benson's views take us into a pre- or postgovernmental structure, a set of worlds on which a future Columbus will land to claim them for himself. Old notions of property rights, embodied in the common-law principles of England and the United States, extended a landowner's realm *usque ad coelum, ad inferos* (all the way to the sky, to the lower depths). When we add hypothetical extraterrestrial civilizations to this concept, we can picture whirling beams of property, driven by the rotation of various inhabited planets, that continuously cross one another, each bearing the label "this is mine"—a *Star Wars* scenario ripe for conflict.

On the other hand, we know of no other civilizations—so far. The notion that the cosmos belongs to those who seize it first appeals deeply to two sets of people, those who feel that we can conquer space only by unleashing the natural human desire for economic gain, and those who would like to throw off the shackles of government altogether, perhaps by creating space colonies that allow individuals free rein. When these two strands combine within the psyche of a single person, their pull becomes immensely attractive. "I want to contribute to humanity getting up and off the planet Earth and exploring the universe," Benson says. "I think [SpaceDev's mission to Nereus] is like breaking the four-minute mile. Once we do this mission, everybody will say, 'Oh, I can do that if he can do that.'... Space is a place, it is not a government program. We can go there; we can make money there."

Benson may be right: Future space exploration may become largely the province of entrepreneurs who went into space seeking profits. One way or another, human society will reach its future, endowed by and weighted with its past and its prejudices. It is to be hoped that we shall learn how to explore space in a reasonable manner and to be respectful of other objects, confining our mistakes, as far as possible, to this our own planet. Whatever we do accomplish will be done by building on the attitudes that our Earth-based society has inculcated in its individual members. In the final pages of his book *Mining the Sky*, John S. Lewis of the University of Arizona, one of the leading proponents of finding raw materials in cosmic objects, describes the future of humanity as follows:

> Once we have a foothold in space, the mass of the asteroid belt will be at our disposal, permitting us to provide for the material needs of a million times as many people as Earth can hold. Solar power can provide all the energy needs of this vast civilization (10,000,000 billion people) from now until the sun expires. Using less than one percent of the helium-3 energy resources of Uranus and Neptune for fusion propulsion, we could send a billion interstellar arks, each containing a billion people, to other stars.... As long as the human population remains as pitifully small as it is today, we shall be severely limited in what we can accomplish. Human intelligence is the key to the future.... Having only one Einstein, one Hokusai, one Mozart, one da Vinci, one Shankara, one Poulenc, one Arthur Ashe, and one Bill Gates is not enough. We need—and can have—a million times as many. We need intelligence, wisdom, compassion, and excellence. These godlike traits are manifested in the physical universe only by life and in the biological universe only by intelligent life. Life is not a cancer of matter; it is matter's transcendence of itself.... The material and energy resources of the solar system allow humankind an infinite future: we can not only break the surly bonds of Earth but break free of the sun and escape its fate. The fulfillment of time and space is matter; the fulfillment of matter is life; the highest fulfillment of life is unbounded intelligence and compassion.

But perhaps it is Lewis who has yet to break the surly bonds of Earth: the naive, all too human arrogance that sees thrusting ourselves through the cosmos as the highest fulfillment of mankind. I'll settle for a human race that survives no longer than the sun's remaining five billion years or so. But what is my opinion in a confrontation with the wealth that lies in the asteroids? I leave it to the reader to decide whether we shall find our highest fulfillment in the resources that await us in space—or in those we can find on the planet where we have been born. When we leave the cradle of Earth, let us go to explore, to learn, and to invent—not to gouge, strip, or despoil the objects of our investigations.

Chapter 7

To Europa And Beyond

For centuries, the planet Mars has dominated human speculations and searches for life elsewhere in the solar system. In 1996 the famed meteorite from Mars gave new impetus to these activities, and Mars remains the prime target for spacecraft that will investigate other planets, hoping to find signs of current or ancient life. Within a few years, we should have automated rovers on Mars, probing for life that might exist in hot springs beneath the surface or clinging to a precarious existence at the fringes of Mars's polar caps.

During the last few years, as bioastronomers (as they like to call themselves) have directed their gaze toward the outer solar system, they have designated Europa, one of Jupiter's four large moons, a rival of Mars for the title of object in the solar system most likely to harbor extraterrestrial life. Though they well knew of Europa's existence, bioastronomers have only recently uncovered evidence propelling it into the forefront of consideration.

Lord of the Planets

Jupiter, largest of the sun's planets, with 11 times Earth's diameter and 318 times its mass, orbits our star at five times the distance from Earth to the sun, or three times the distance from the sun to Mars. A retinue of nearly twenty moons accompanies Jupiter in its orbit. Most of these satellites, only a few dozen miles across, are probably former asteroids captured billions of years ago from the well-populated asteroid belt between Mars and Jupiter. But four of Jupiter's moons are exceptionally large, with diameters equal to or much greater than that of our own moon. The nearly circular orbits of these four large satellites suggest that they formed in orbit around the largest planet during the era when Jupiter pulled itself together, 4.5 billion years ago.

Like Saturn, Uranus, and Neptune, Jupiter consists of enormous amounts of hydrogen and helium gas surrounding a solid central core. In addition to the dominant hydrogen and helium, Jupiter's gaseous outer layers contain significant amounts of carbon monoxide, carbon dioxide, methane, ammonia, and water vapor. These are the simplest compounds that hydrogen forms with the elements carbon, nitrogen, and oxygen. The mixture of molecules in Jupiter's outer regions resembles the composition of Earth's atmosphere billions of years ago, at the time when life originated on this planet. However, in all of Jupiter's thousands of miles of gas, the planet lacks what seems to be an important element, perhaps a requirement, for the appearance of life, namely liquid pools or solid surfaces where complex molecules can collect and interact.

Jupiter's Galilean Satellites

In contrast to the gaseous outer layers of Jupiter itself, the planet's four large moons are solid all the way through. These objects have too little mass to have prevented gaseous hydrogen and helium, the two lightest elements, from escaping during the billions of years since they formed. The four large moons consist of minerals and simple molecules made from the list of the most abundant elements in the solar system past hydrogen and helium, including silicon, oxygen, aluminum, magnesium, carbon, and nitrogen—the same elements that form the bulk of Earth and the sun's other inner planets, Mercury, Venus, and Mars. In 1609 the Italian astronomer Galileo Galilei discovered these four moons when he turned his telescope on Jupiter; he called them the "Medicean stars" in an attempt to secure further funding from his patrons, the Medici family. Other astronomers rejected this rather crass nomenclature and named the moons after four of Jupiter's mythological lovers: Io, Europa, Ganymede, and Callisto. In compensation, however, the four large moons became collectively known as Jupiter's Galilean satellites after their discoverer. For centuries, astronomers observed these points of light orbiting the giant planet and eventually used improved telescopes to discover other, smaller moons.

We knew next to nothing—only their approximate sizes and masses—about the four Galilean satellites until we sent spacecraft past Jupiter. Two *Pioneer* and two *Voyager* spacecraft sailed by the Jovian system during the 1970s, obtaining the first close-up images of the planet and its moons. These photographs made one of the Galilean satellites famous: Io (pronounced either EYE-oh or

EE-oh), which has a multicolored, pizza-like appearance. On Io, volcanic eruptions spew sulfur-laden gases dozens of miles high, providing the moon with a tenuous, constantly renewed atmosphere of foul-smelling, unbreathable compounds that soon fall to the surface, continually recoating it with a new set of variegated colors.

Though Io's volcanoes are spectacular and have something in common with early Earth, where volcanoes likewise once threw new types of molecules into the primitive atmosphere, they offer little hope that life might exist on Io today. The continual rain of hot sulfur imposes a hellish fate on any complex molecules that might form, breaking them apart before they can produce the long-chain molecules typical of life. Io's photogenic qualities are antilife ones, and we must look elsewhere if we hope to find extraterrestrial life in the solar system.

Europa, the Darling of Ice

In our quest for extraterrestrial life, the Galilean satellite that has gained the spotlight is Jupiter's second large moon, Europa, named after the goddess who gave her name to the subcontinent of Europe. With about the same size as Io and our own moon, Europa has a dull, nearly colorless surface whose appearance and light-reflecting properties are strongly suggestive of a solid quite familiar to us on Earth, frozen water. Europa's surface, first photographed by the *Voyager* mission, received special attention from the *Galileo* spacecraft that reached Jupiter at the end of 1995 and began to execute a series of long, looping trajectories carrying it successively past each of the Galilean satellites. The *Galileo* photographs have convinced almost all experts that Europa indeed has a covering of water ice—not frozen carbon dioxide, which forms much of the Martian polar caps, or another frozen substance, but familiar water, frozen at temperatures hundreds of degrees below zero Fahrenheit, far colder than even the frozen continent of Antarctica.

Ice is nice, but a frozen world by itself offers few attractions. Europa's claim to fame rests not on its ice but on what may lie below the ice: a worldwide ocean of liquid water. The ice that the *Galileo* orbiter has photographed may well be floating, like the Arctic ice pack on Earth, on slightly denser water below. If Earth had no pronounced topographical features and were as cold as the Arctic regions, all of our planet would possess a similar icy surface with water underneath it.

If Europa does have a worldwide ocean beneath its ice, it qualifies as the only solar system object except Earth where significant amounts of liquid water exist. (The *Galileo* spacecraft has recently obtained tantalizing, indirect hints that Callisto, the outermost of Jupiter's four large satellites, may also have water beneath a worldwide layer of ice, but this remains largely speculative.) Because water, or another substance that can remain fluid, ranks in most assessments as the number-one requirement for life, the thought of abundant water on Europa acts upon bioastronomers like catnip on Sylvester. The possible, even probable, ocean beneath the ice of Europa could be teeming with extraterrestrial life forms crying out for investigation and comparison with life on Earth. All that separates us from certifying this conclusion is verification of the assertion that Europa's ice indeed conceals an ocean, plus a journey to Europa with equipment that could probe beneath the ice and search for any forms of life that may float or swim there and even now find themselves on the path to eventual evolution into intelligent creatures.

Perhaps we are getting a bit ahead of ourselves here. We have gained a knowledge of Europa's icy crust but must pause to ask, How and why could this moon, at five times Earth's distance from the sun and consequently far colder than anywhere on Earth, maintain water in its liquid state? And if it could, why should we expect that the entire satellite has an ocean of liquid water beneath its icy exterior?

The first of these entirely reasonable questions has a believable answer. Europa can in theory maintain an ocean beneath its ice cover because radioactive rocks in its interior release enough heat to keep this hypothetical ocean from freezing. The production of heat by radioactive minerals explains why the temperature rises as we descend through Earth's crust (a phenomenon that makes mines that plunge several miles beneath the surface almost unbearably hot). Inside Europa, which has only about one-quarter of Earth's diameter, heating by radioactive rocks produces a much lesser effect but one that calculations show to be sufficient to keep water liquid—provided the water has a thick covering of ice. The ice inhibits the loss of heat from the water just as the Arctic ice pack does, allowing the water below to remain liquid.

Thanks to its radioactivity, Europa may indeed possess water beneath its ice. But does it? The evidence for a worldwide ocean on Europa rests largely on the *Galileo* photographs of the moon, which show continual subtle changes in the positions of the blocks forming its surface. Over a period of several days or weeks, these blocks move with respect to one another, implying that they are floating or gliding on a deeper layer that supports them. This suggests but does not demonstrate that Europa's ice forms an overlying layer above an ocean. Just how thick the ice is and how deep the ocean is remain unknown. Estimates of the ice layer thickness range from a few yards to half a mile or more, and if the ice floats on water, the water might be anywhere from a few yards to several miles deep.

Cracking Europa's Crust

How can we improve our knowledge of Jupiter's most tantalizing satellite and discover whether or not it possesses the prime qualification for life, liquid water? The answer lies with expeditions to Europa, first with automated spacecraft and then with human explorers. The primary goal of these expeditions will be to find a way to pierce the layer of ice, first probing for evidence of a watery environment and then, if a Europan ocean does exist, going beneath the ice to explore this strange world.

Even the first of these tasks seems daunting. How can we plan to break through an icy layer whose thickness we do not know? If we find that Europa's ice layer extends downward for, say, half a mile, how can we pass through such a thick crust of ice? What equipment can we bring to Europa to explore a subsurface ocean, supposing that one exists?

Difficult problems call for determined efforts. To break through a thick layer of ice, we could employ the relatively complex method of drilling or the much simpler approach of bombing. Drilling through the ice would require sending equipment to Europa similar to that used to drill oil wells on Earth but capable of operating at temperatures close to -250 degrees Fahrenheit. This equipment could be automated, assuming that we could master the logistical problems involved in manufacturing a rig that can operate in Europan cold, packaging it into a spacecraft, sending it about half a billion miles farther from the sun than our planet, and operating it by remote control, when radio signals take more than half an hour to pass between ourselves and Europa. Bombing could be far more easily arranged: We would send what amounts to a missile to Europa and detonate its warhead close to the surface. The explosion should have a dual result, both exposing what lies beneath the ice (provided of course that we have packed a sufficient charge to blow a hole through ice that may be half a mile thick) and expelling debris that would include bits of whatever lurks beneath Europa's frozen surface. A spacecraft trailing the bomb run could then not only photograph the results on Europa but also scoop up debris and test it for signs of life.

The physicist Freeman Dyson has pointed out that this sort of explosion merely mimics what must occur naturally from time to time when a large rock, like a meteoroid in our vicinity, strikes Europa. The highly cratered surfaces of Callisto and Ganymede, two of Jupiter's other large moons, have received such large impacts, which must have repeatedly broken the crust of Europa and penetrated deeply into any ocean that may lie below. In that case, each impact should have shot a spray of water and debris into the region occupied by Jupiter and its moons. Some of this debris, captured by Jupiter's immense force of gravity, should be in orbit around Jupiter now; a ring of small objects orbits the planet, but we have no way of knowing where they came from. Dyson therefore suggests sending a spacecraft into the debris belt to search for freeze-dried fish from Europa's ocean! Here Dyson makes a joke, but only by exaggeration: If Europa does have an ocean with solid material floating in it, some of that material ought to have been expelled by impacts, and part of it should still be in the ring of material orbiting the largest planet.

Creating another impact would allow us to examine the stream of material from Europa itself rather than trying to sort out which debris in Jupiter's ring came from Europa. To be sure (and here we joke along with Dyson), depending on just what might be lurking beneath the surface of the ocean, our Europa bombing mission could be interpreted as hostile, because most of us would resent a similar mission from a faraway world sent to examine Earth. We can take some comfort in a practical if not a moral dimension by concluding that whatever may live beneath the ice of Europa (and there may be nothing at all), it—or they—have apparently done nothing to alter the appearance of Europa's surface: If the Europans have weapons of mass destruction, they have yet to deploy them on the ice.

From a scientific viewpoint, the sadder outcome would be that any drastic interference with the Europan environment (and doubtless some less drastic ones as well) would risk altering the situation forever, so that later explorations would be unable to determine what the pre-human investigation conditions actually were. As in the case of Mars—or any celestial object that we hope to probe for extraterrestrial life—we must proceed carefully if we hope to avoid terrestrial contamination of another world's environment. Bombing Europa provides a "cheap and dirty" way to study this large moon, but one that we might come to regret before too many generations had passed.

The Drilling Expedition

Suppose, then, that we do the right thing and send a drilling mission to Europa, sterilized to the maximum extent that we can arrange. Extrapolating from our current abilities and the speculations we constructed in the last chapter, we can estimate that an automated drill team could be in place by the fourth decade of the twenty-first century, and a human mission could be ready to embark a few decades afterward. The trip would not be overwhelmingly long—just about four times longer than the journey to Mars, hence about four times longer in time if the same propulsion mechanisms were used for both journeys. In other words, we are talking about a voyage that would last for two or three years, easily accomplished by automated spacecraft. The success of the *Pioneer, Voyager,* and *Galileo* missions to Jupiter (and, in the case of *Voyager 2,* to Saturn, Uranus, and Neptune as well) testify to our ability to construct robots that can easily survive a dozen years in space. With a sufficient investment of time and engineering, we could create automated explorers that would land on Europa, drill through its ice (even ice half a mile thick), and lower instruments into the dark ocean below.

Humans, of course, could do still more. Even though they require special attention, once we surmount the problems of sending a human mission to Mars, we could doubtless try a bit harder, and find astronauts a bit hardier, to make the longer journey to Jupiter and its moons. Picture, then, roughneck drillers on Europa about a century from now—as distant in time from our present era as we are from the heroes who built the Panama Canal—who land on the ice and then build an outpost to maintain the machinery that will drill through the ice to the ocean below. This image presupposes that automated explorations of Europa have already determined the thickness of the ice and that the ice does float atop water. The astronauts would therefore have brought the right amount of equipment for the drilling process, along with probes to be lowered through the ice to drift through the sub-Europan sea. Once again, automated investigations should have already determined whether or not life exists in this ocean; if the answer is yes, the primary goal of the first human expedition would be to find out more about the chemical composition, metabolic activity, and evolutionary state of as many life forms as possible. We can imagine that the astronauts have brought a robot submarine to sail in the Europan sea, sending images and data to the surface from which they can be sent to Earth. Eventually, properly supplied from Earth or from a spacecraft in orbit around Jupiter, humans themselves might descend through the ice of Europa to examine whatever lives below, the product, like ourselves, of four billion years of evolution in a solitary world in the solar system.

The Lakes Beneath Smog-Shrouded Titan

Though Europa, possibly in a tie with Mars, now appears to be the object within the sun's family most likely to harbor extraterrestrial life, another large moon in the outer solar system commands our attention: Titan, Saturn's largest satellite. With one and a half times our moon's diameter, Titan has more than three times the volume of Io, Europa, or our moon, and misses being the largest moon in the solar system by only about a hundred miles. Ganymede, Jupiter's largest satellite, shows us a battered, rocky surface with no evidence of an atmosphere, oceans, or any suggestions that life may once have existed amid its craters and ridges. But Titan, on the other hand, has a different tale to tell us, one of mystery, intrigue, and a rich potential for life.

Titan is unique among all solar system moons because of its thick atmosphere, which consists mainly of nitrogen molecules, the same molecules that form the bulk of the atmosphere on Earth. In fact, the atmospheric pressure on Titan's surface roughly equals that on Earth. The pressure and composition of the atmosphere suggest that an astronaut standing on Titan's surface would find herself in the most Earth-like environment in the solar system (not counting Earth itself).

However, this environment can be considered Earth-like only with an elastic use of "like." Orbiting at 9.5 times Earth's distance from the sun, Titan remains in a celestial deep freeze, with a surface temperature of less than -300 degrees Fahrenheit. And even though Titan's atmosphere duplicates the bulk of Earth's, Titan lacks what we consider an essential part of our air, the oxygen molecules that form 21 percent of Earth's atmosphere and allow animal life to exist. A nitrogen-rich atmosphere does not by itself imply the origin of life.

On the other hand, plants and algae can live perfectly well on Earth without oxygen, though they do require carbon dioxide, which Titan possesses in extremely modest amounts. We should attempt to broaden our horizons when we consider the possibilities of extraterrestrial life, using life on Earth only as a guide, not a template. The broadest specifications for life requires some liquid, a fluid within which molecules can float and interact, steadily producing more complex molecules that may eventually lead to living organisms. In the case of Europa, the potential fluid for life appears to be water, playing the same role there that it does on Earth. On Titan, which is far too cold for liquid water to exist, a good chance exists that life's liquid can be provided by another fluid: liquid ethane, a hydrocarbon molecule made from hydrogen and carbon atoms.

Hydrocarbons are well known on Earth, where we use some of them (most notably methane, ethane, and propane) as fuel. All these hydrocarbons react strongly with oxygen; if the reaction proceeds quickly, we call it "burning" the hydrocarbons. As a result, we cannot and do not expect to find large pools of hydrocarbons on Earth's surface because exposure to oxygen in the atmosphere would sooner or later consume them. On Titan, however, which has almost no oxygen in its atmosphere, hydrocarbons can persist for billions of years. Because some hydrocarbons—ethane most notably—can remain liquid even at extremely low temperatures, we can imagine that Saturn's largest moon has lakes or even seas of ethane, perhaps with methane and other hydrocarbons mixed in.

Imagining these lakes describes our present state of knowledge because no one has observed Titan's surface, which may yet prove to be dry as the Sahara Desert. Titan's opaque atmosphere always blocks our view, interposing a haze so thick that Los Angeles smog seems mild by comparison. Like the smog in California, Titan's atmospheric opacity arises from the presence of long-chain molecules that form when sunlight shines on relatively rare types of molecules in the atmosphere; in Titan's case, these molecules are mainly hydrogen cyanide. As some of the smog falls onto the surface, sunlight continually makes new long-chain molecules, and so Titan's atmosphere never grows clear. The surface of Titan never sees the sun directly, and the amount of light that leaks through the atmosphere (much reduced to begin with, because Titan orbits at nearly ten times the Earth-sun distance) falls far below familiar levels on Earth.

Nevertheless, Titan may have pools of liquid and certainly possesses the raw materials for life; in addition to nitrogen and hydrogen cyanide (not the best start toward life), its atmosphere contains carbon monoxide, carbon dioxide, a tiny amount of water vapor, and various hydrocarbons. If Titan does have liquid pools or seas, it may well also have a complete cycle of evaporation and rainfall, as ethane rainstorms and waterfalls constantly change the surface and furnish microenvironments analogous to the tidepools in which life may have begun on Earth.

The Exploration of Titan

All that remains is to visit Titan and see what the facts are. Automated spacecraft have begun this task, first with the *Pioneer* and *Voyager* flybys of the giant satellite in the 1970s and then with the far more advanced *Cassini-Huygens* spacecraft, launched from Earth in 1997 and scheduled to reach the Saturnian system in 2004. The *Cassini-Huygens* mission includes an orbiter that will make detailed observations of Saturn, plus a separate spacecraft, called the *Huygens* probe (after Christiaan Huygens, the Dutch astronomer who discovered Titan three centuries ago) that will descend through Titan's smoggy atmosphere, deploy a parachute, and land on Titan's surface. The *Huygens* probe may not be able to survive for long in the cold (and wet?) environment in which it will find itself, but during its active lifetime it should send images of the surface back to Earth along with measurements of the composition of the surface and the atmosphere just above. In addition, the *Huygens* probe will determine the density, temperature, heat conductivity, and electrical conductivity of the lower atmosphere and will indicate whether it has encountered a liquid or a solid environment.

The *Cassini-Huygens* mission to Saturn ranks as the most impressive that the United States, in collaboration with its European friends, has launched into the outer solar system. As such, the mission can serve as a benchmark of what we can do now and what we must improve on if we hope to look farther and with greater precision into the cosmos that surrounds us. The *Cassini-Huygens* spacecraft weighs in at just over six tons, almost as much as the two probes that the now-defunct Soviet Union sent to Mars. (Perhaps fittingly, the two probes never functioned any better, and in fact a good deal worse, than the Soviet Union itself). Of the six-ton total, the *Cassini* orbiter weighs about two and one-third tons and the *Huygens* probe about a third of a ton; the remaining three tons consist mainly of the fuel that will allow the *Cassini* orbiter to change its trajectory into an orbit around Saturn and then vary that orbit to pass close by many of Saturn's large moons.

For truly long-distance travel, saving on weight will become critical, and the propellant would be the first to go—if we can find a better way to allow the spacecraft to maneuver. When it comes to generating power for its own needs (other than maneuvering in flight), the *Cassini-Huygens* spacecraft, like *Galileo* before it, makes do with a small amount of radioactive plutonium, only a few pounds in weight, whose decay can provide sufficient energy to run the spacecraft's instruments for many decades. A problem arises from the high toxicity of this plutonium, as well as its radioactivity; if these few pounds of plutonium were spread through Earth's atmosphere, large numbers of people would become gravely ill, and some would die. This issue provides more than academic interest because both the *Galileo* and *Cassini-Huygens* spacecraft use Earth's gravity to boost their speed, allowing them to proceed to the outer solar system. To do so, they make a close pass by Earth, and in the case of the *Cassini-Huygens* spacecraft, which has already passed by Venus for similar purposes, that pass will not occur until August 1999. NASA has calculated that only a minuscule chance exists that something could go so wrong that the *Cassini-Huygens* spacecraft would enter Earth's atmosphere and be consumed by friction; additional reassurance comes from the fact that the cake of plutonium would probably survive passage through the atmosphere as a single lump, and be drowned (more likely than not) in miles of deep ocean. On the other hand, sometimes events with only a tiny chance of occurring actually do happen.

When we look to the future and ask how we shall power our long-ranging spacecraft, the attractiveness of obtaining energy from the decay of radioactive nuclei such as plutonium must be balanced against the risks, not only to Earth (if we launch from Earth or use Earth's gravity for an early boost in velocity) but also to the astronauts who ride the rocket and stay with the spacecraft for years. The danger from radioactive and toxic elements used as a power supply counts as one more argument against sending humans on dangerous missions—always to be overcome, if need be, by securing volunteers who find that the excitement far outweighs the dangers. From a preservationist standpoint, the gravest danger from spacecraft that use radioactive nuclei to produce their energy, or from installing larger nuclear power plants on other objects in the solar system, is the possibility that an accident could spew radioactive material over the surface of another world, rendering further exploration dangerous if not impossible and damaging any forms of life that might already exist.

The next few centuries should allow us to make a complete close-up survey of the entire solar system, certainly with automated instruments and quite possibly with human explorers. As we complete and refine these investigations of the worlds bound to the sun, interest should grow in the next great step in exploring our surroundings: travel to planetary systems far beyond the sun's family. Before we examine futuristic propulsion systems that could someday make this dream a reality, we should pause for a proper introduction to our starry galaxy, whose dimensions may awe us as thoroughly as its possibilities inspire us. Let us meet our stellar home, the Milky Way.

Chapter 8

A Million Times Farther

Far beyond Pluto, the sun's outermost planet, far beyond the billions of comets that orbit the sun, far, far indeed lies the rest of the Milky Way galaxy.

How Far Are the Stars?

The plain fact about astronomy is that its facts are too arcane for us to grasp intuitively. This truism becomes evident from the most basic description of the arrangement of objects in the cosmos. Because it violates our intuition, humans took a long time to admit that Earth does not form the center of the universe. For the same reason, as well as some technical difficulties, we have found it difficult to accept the fact that the distances between the sun's planets are hundreds or thousands of times greater than the size of our own planet—not thousands of miles but tens of millions, hundreds of millions, or even many billion miles. Though our tongues can easily recite these numbers, our inner selves rebel. Why should we believe that a scale model of the solar system, made with an orange to represent Earth, would put Mars somewhere between a quarter of a mile and two miles away, depending on the planets' positions in the orbits? Or that in this model, the sun, as large as a small house, would be half a mile from Earth? And that Jupiter would be two and a half miles away, Saturn five miles, and Neptune twenty miles from Earth? What use is all that space? How unlikely, how far from human needs!

By dint of careful repetition, much of the educated population has achieved conscious mastery, at least in moments of mental application, of the immense size of the solar system. But for almost all of us, such suppression of our inner child's belief that space cannot be this large represents our maximal effort. The next step, recognition of the fact that the immense distances between stars make the solar system tiny by comparison, is one step too far, one "immense" more than we can absorb.

Yet these are the facts, prized by the astronomers who deduced them (not without a long struggle over abandonment of the belief in a privileged position for the solar system) and available almost without seeking to anyone who watches educational television or searches out the most isolated section of a bookstore to page through the science offerings. In thinking about the distances between the stars, the number to remember is one million, about equal to the number of dollars that an average employee will earn in a lifetime, or the number of people who live in greater St. Louis. One million also approximately equals the factor by which the distances to the stars closest to the sun exceeds the distance between the Earth and the other planets in the solar system. Not by a thousand times, or ten thousand, or a hundred thousand, but a million! There is no use nodding in agreement, for our inner psyches will never accept such a concept. Yet there are the stars, silently shining from distances so great that they make our heads spin.

Though we may—and shall—add details, the reader might better look to his or her conception of the cosmos, taking care not to assent too quickly. Having spent years in training as an astronomer, I believe that the sun lies ninety-three million miles from Earth and that the sun's closest neighboring stars are about twenty-five trillion miles away. I know it, believe it, and live with it every day—but there is a part of me just as certain that the stars are only an immensely great distance from us, and an "immensely great distance" equals a few tens of thousands of miles. That part of me has a highly limited vocabulary and barely bothers to talk to my highly developed consciousness, but when it mutters its few, indistinct sounds, the rest of me seems to listen carefully. In short, knowledge is power, and the less intellectual part of us has no plan to surrender its power by abandoning what it considers to be knowledge. Since this part of our psyches lies beyond intellectual persuasion, I cannot hope to convince it; the best I can do, ironically enough, is to reinforce my knowledge of the truth by repeating it like a mantra.

Measuring by Light

When astronomers discuss the distances between stars, they often employ a distance unit called the light-year. In fact, they more often use a still more advanced unit called the parsec, but this merely testifies to the fact that each specialty has reasons to love its own jargon; for us, mere light-years will suffice. One light-year equals the distance that light travels in a year; it therefore specifies a distance, not an amount of time, as the word "year" implies. Picture a beam of light arrowing through space, traveling, as light does, at a speed close to 186 thousand miles per second. Let 31.6 million seconds elapse, and the light will have traveled about 6 trillion miles. Thus the number of miles that light covers in a year roughly equals the number of dollars that pass through the hands and bank accounts of all the wage earners in the United States every three years. In a comparison that makes a more direct reference to time, 6 trillion also equals the number of seconds in 186 thousand years (Can you see why this is so?), or about 2.5 thousand human lifetimes. Not to put too fine a point on it, 6 trillion is a number so large that it falls, even with the best references and intentions, into the category of "really large numbers."

Yet there the number stands: One light-year—6 trillion miles—takes us less than one-quarter of the way to the sun's closest neighbors, which lie 4.4 light-years from the solar system. We need not bother to specify which part of the solar system we are using as our benchmark, since Pluto's orbit, forty times as large as Earth's, spans a distance less than one-thousandth of a light-year. Sunlight, which takes eight minutes to reach us from our star (hence we would enjoy eight minutes of happy ignorance if the sun should suddenly disappear), requires more than five hours to reach Pluto, whose distance from us and the sun is therefore about five light hours. Plunging outward past Pluto at its enormous velocity, sunlight must travel not for days, weeks, or months but for several years at least before it encounters any objects familiar to us. In a model with Earth as an orange and the sun as a cottage half a mile away, the closest stars are represented by other cottages—halfway to the moon.

Living in the Milky Way

For centuries, the arrangement of the cosmos beyond our solar system remained largely a mystery. Despite their best attempts at counting the numbers of stars visible in different directions, astronomers failed to perceive the actual distribution of the stars surrounding our sun. This failure arose in part from limitations in the ability of astronomers to view the cosmos, but mostly because dust particles floating among the stars block the passage of starlight. These dust particles block light in a nonuniform way because they are unevenly distributed through interstellar space. As a result, when we look outward from the solar system, some of our lines of sight reach outward almost indefinitely while others extend only to astronomically short distances before the opacity imposed by interstellar dust prevents us from seeing farther.

This uneven effect keeps us from seeing a cosmic truth that would otherwise shine strikingly on every clear night. Our solar system resides within a mammoth system of stars shaped like a thin plate with a bulge at its center. We inhabit a region far from this central bulge, well out toward the outer edge of the distribution of stars. If we had a clear view of the cosmos in all directions, we would see a circle of light shining in the sky, the light from the total distribution of stars concentrated into a thin plane containing the solar system. Half of this circle would always be above the horizon, while the other half remained below. One point on this circle would appear particularly bright with light from the multitude of stars that form the central bulge in the stellar distribution.

Because of interstellar dust, this radiance appears in a highly muted form. We can and do see a circle of light around the sky, the diffuse band that we call the milky way, which indeed arises from the combined light of a host of individual stars. We can also see the light from the central bulge at one point along the milky way, but it shines only a bit brighter than the rest of the circle because interstellar dust has a particularly high concentration in its direction. If we look closely at the milky way, for example, as it runs through the constellation Cygnus (visible throughout the summer and fall), we note that the band of light seems to split into several starry streams, as if the milky way has divided itself. This is an illusion: We are seeing a blockage of light by interstellar dust, which creates dark lanes amid the light of the milky way. When we observe these dark lanes, our eyes and brains tell us that we are looking outward toward greater distances, when in fact dust grains terminate our lines of sight by blocking starlight that would otherwise add to the milky way.

More than half a century ago, astronomers finally recognized the importance of interstellar dust. They realized that the sun lies within the flattened distribution described above, an assemblage of stars shaped like a large, flat tea tray with a grapefruit through its center. Only the closest stars in this assemblage appear to the unaided eye as single points of light to the unaided eye; the more distant ones concentrate in the circle of starlight long called the milky way. Using capital letters to create an important distinction, astronomers introduced "the Milky Way" as the name of the collection of stars to which the sun belongs. Eventually, they established that the Milky Way galaxy contains about three hundred billion stars and spans about one hundred thousand light-years from rim to rim. The solar system lies approximately thirty thousand light-years from the central bulge of stars, hence well out in the suburbs.

The word "galaxy" is derived from the Greek word for "milky," and astronomers use it to describe the tremendous agglomerations, each containing millions or billions of stars, that illuminate the universe. The starry universe is based on billions of billions. Within ten or twelve billion light-years of the Milky Way, the maximum distance to which we can observe, the cosmos contains billions of giant galaxies comparable to our own. Each of these galaxies has hundreds of billions of individual stars, and simple multiplication shows that the visible universe shines with the light from about 10,000,000,000,000,000 stars—a number sufficiently large to boggle the mind, and one that agrees with the notion that all astronomically large numbers are the same.

For astronomers, the nearest of these stars are familiar individuals, with their starlight quirks as well known as the drive to work. The sun's closest neighbors in the Milky Way are the three stars that form the Alpha Centauri system, 4.4 light-years away. When we seek to comprehend these light-years and to use them in imagining the full facts of the Milky Way, we note that Alpha Centauri's distance is about ten thousand times the diameter of the solar system and one ten-thousandth of the distance to the galactic center. Since a trip to Mars would take us a distance about one-hundredth of the diameter of the solar system, the preceding numbers imply that a voyage to Alpha Centauri would take a million times longer than a journey to Mars, while the galactic center lies fully one hundred million times farther from us than our planetary neighbor does.

Implications of the Enormous Size of the Milky Way

These enormous ratios of distances do not of course imply that the journeys are impossible. We devote the next few chapters to examining futuristic spaceships capable of interstellar voyages and to shortcuts and possibilities that make some of the problems associated with these voyages potentially less imposing than they appear to be from our present vantage point, when sending humans to Mars lies at the edge of our capabilities. Before we embark on these considerations, however, we would do well to review the implications of the stellar distances and numbers that describe the Milky Way.

First of all, our solar system belongs to and forms part of the paradise we call the Milky Way. The notion of a voyage *to* the Milky Way must therefore yield to the concept of a journey *through* the Milky Way. Even though every trip through the solar system counts as an extremely modest voyage through our galaxy, we can rightly reserve this phrase and concept for journeys that cover interstellar distances, taking us at least halfway to the closest stars.

Second, three hundred billion stars in the Milky Way imply that our experiences on Earth quite likely have counterparts elsewhere. As we shall discuss in detail in Chapter 11, the galaxy's vast number of stars, most with presumed planetary systems, leads to the reasonable (but questionable) conclusion that a voyage through the Milky Way could easily take us past a large number of sites with their own highly developed civilizations. Thus a voyage of discovery might well turn out to be a chance to meet beings who know a good deal more about the galaxy than we do; in other words, our travels might well lead us to the results of previous explorations, surely a bonus in addition to the thrill of establishing contact with another civilization.

Third, the splendid prospects of interstellar space travel lie many generations into the human future and will themselves require many more generations of travel time. We cannot now determine of course how much time will pass before humans find their way to other star systems in the Milky Way, how they will do so, or what they will find on these journeys. What, then, can we be sure of as we contemplate a future so glorious that we may be able to travel for significant distances through our Milky Way?

How about this: We can be sure that before humanity develops the technology necessary for interstellar space journeys, our society will have undergone such immense changes that we would have difficulty recognizing it. In particular, the accumulated knowledge that has allowed us to survive and to endure on Earth may or may not prove capable of maintaining us in a space environment. As we stand on the threshold of space exploration, we err by assuming that we can confidently decide now whether or not our adaptive skills will eventually prove capable of discovering how to embark on interstellar spaceflight and how to colonize worlds so distant from the solar system that the colonists may have no plans to return, or even to stay in touch with the home planet. In short, we should admit our uncertainty, even though our emotions tend to tug us toward one conclusion or its opposite, that the human spirit can overcome all obstacles or that we are fated never to penetrate the immense vastness of space in the Milky Way.

Long before these journeys become reality, we may establish contact with other civilizations by radio or even by visits to our small planet, as discussed in Chapter 11. If this does not happen, we are likely to send automated probes long before we risk human lives on these exciting but dangerous voyages of exploration—probes that can travel for millennia without the demands for food, drink, and air that characterize humans. To construct even these probes requires a technology far more advanced than our own, plus a willingness to invest in truly long-term projects. However, if we want to understand the prospects as well as the hazards, we must turn to the scientific and technological issues raised by our desire to visit the other stars of the Milky Way. Let us turn to a bit of rocket science to see what, if anything, limits our dreams.

Chapter 9

Rocket Science, Laser Sails, and Antimatter Drives

The cosmic arrangement of matter has set stars so far apart that journeys to neighboring planetary systems must cover enormous distances, so vast that in contemplating them the human mind tends to reel, unable to assimilate the number of times that the distance to other star systems exceeds the already huge distances that describe voyages within the solar system.

But these are the facts of life in the Milky Way. If we hope to travel to other stars and their planets, we must confront the fact that most of them lie millions of times farther from us than Mars or Venus. Since humans have an impressive track record in attacking technological problems, we can reasonably assume that with sufficient time and insight, we may well find a way to make not just interplanetary but also interstellar journeys, covering not the hundreds of millions or billions of miles to the planets but rather *millions* of billions of miles—distances so great that measuring them in miles is only confusing. By using units of light-years, we can simplify the numbers: The closest stars are a few light-years from the sun, and a million stars lie within a thousand light-years. Because a single light-year embraces a distance about a thousand times greater than the largest planetary orbit in the solar system, we cannot expect to create an interstellar spaceship overnight. Nevertheless, the time may come when humans embark on their first journeys to the stars.

What Do We Need for an Interstellar Voyage?

The fundamental requirements for a trip to other star systems match those for any travel within the solar system, except that quite obviously we shall need more of everything for a journey that lasts much longer. Our basic needs will remain the big five: fuel, food, water, air, and protection against harmful radiation. To this list we might also add time, by which we mean that if an interstellar voyage will last for centuries or millennia, either we must develop serious life extension mechanisms so that explorers can live for these time intervals, or else we must establish a social order aboard the spaceship that will keep succeeding generations devoted to their task, even though they will see neither the start nor the end of the journey. Strangely enough, if we allow ourselves full freedom to speculate, the problem of time might find an easy solution in voyages that take no more than a few decades to cover interstellar distances (see page 197).

To earn this speculative freedom, let us take a good look at the most basic problem of all, finding the fuel for the journey, the source of energy that will send our spacecraft across the light-years. By our definition, a "good" source of energy not only will propel a well-shielded spaceship but will also allow its crew to produce supplies of food, water, and air from the raw materials we take with us or encounter along the way. In other words, the engine that drives the spacecraft should also furnish the power for activities on board, including raising crops and refining water and oxygen from material collected along the journey. If we want to see how we can satisfy our needs for an interstellar space voyage, we must discuss a bit of rocket science.

Four Types of Spacecraft Engines

How can we impel an interstellar spacecraft to cross trillions of miles and reach another star system? One possible propulsion system leaves the engine at home. Instead of carrying a rocket motor, the spaceship deploys wide sails, which catch not the wind but the blast from a powerful laser beam carefully directed toward the sails while avoiding the crew inside the ship. By not requiring that the spaceship carry an engine or its fuel, this arrangement reduces the size and weight of the craft considerably.

Laser sail propulsion, first proposed by the physicist Arthur Kantrowitz three decades ago, has become the special project of Robert Forward, an engineer better known for his science fiction novels. Forward has conceived of a laser-driven starship with sails many miles across—not an easy vehicle to build, but the job is made notably less difficult by the prospect of constructing the starship in space, free of the distorting effects of gravitational forces. Even larger than the sails, though not required to travel through interstellar space, would be the lens that focuses the laser beam. Forward envisions creating the lens, made from concentric, alternating light and dark rings of extremely thin plastic, between the orbits of Saturn and

Uranus. The lens, as large as the state of Texas, would be continuously kept in exactly the right orientation to focus sunlight onto the spacecraft as it speeds away from the solar system.

Why do we call this focused light a laser beam? A laser produces a highly directed beam of light of a single, precise color made of massless particles called photons. Each photon carries some energy, and the different colors of light consist of and are defined by photons that carry different amounts of energy. The lens that Forward proposes would focus light of a single color into a highly "collimated," extremely narrow beam that could be directed precisely toward the sails of an interstellar spacecraft. By pushing on the sails for a year or so, the laser beam could accelerate the spaceship to a fair fraction of the speed of light—a speed that would allow it to reach the closest stars within one or two decades.

By leaving its motive power behind, a laser sail spaceship can save a tremendous amount in its mass requirements, though it must still of course carry sufficient supplies of food, water, and oxygen for its crew. This spacecraft does, however, suffer from two severe handicaps. First, a laser beam always transfers its energy to the sails of a spaceship in an inherently inefficient manner, which implies that the laser propelling the spacecraft must be even more powerful than we might imagine. Second, as the spaceship journeys ever farther from Earth, the impulse from the laser beam grows ever smaller because even the tightest beam inevitably spreads out, covering a much larger area than the target offered by the sails. The part of the beam that misses its target simply goes to waste, as far as propulsion is concerned, and an ever-decreasing portion of the laser's total output helps push the spacecraft. The need to focus the beam leads to the requirement of a lens as large as Texas, even for a trip to the closest stars to the sun; for longer journeys, we would need significantly larger lenses.

Let us not abandon the laser sail concept prematurely. Remember that a spacecraft doesn't require the continuous application of a force to keep it moving. Instead, once the spaceship has been accelerated to a given velocity, it sails through space at that velocity as long as it remains so far from stars and other massive objects that they exert only insignificant gravitational forces on it. Thus the laser sail approach to interstellar voyages demands only that the laser beam accelerate the spaceship to the speed it will maintain for its journey. The downside of this concept lies in the low efficiency of the acceleration process, which implies that we must apply the laser beam for a long period of time during which the spacecraft moves farther and farther from Earth.

An intriguing variant of the laser sail spacecraft foresees the use of power from microwaves rather than a laser beam. Microwaves are one type of electromagnetic radiation, characterized by relatively long wavelengths and low frequencies of the vibrating photons that constitute the radiation. Radio waves have the longest wavelengths and lowest frequencies of all types of radiation, but next come microwaves, which have proven useful for cooking because they have wavelengths and frequencies that penetrate food, agitating it to the point at which we find it palatable. An American physicist and a Russian physicist, Leik Myrabo of Rensselaer Polytechnic Institute and Yuri Raizer of Moscow's Institute for Problems in Mechanics, have collaborated in studying possible "microwave lightcraft" that would ride a beam of microwaves to reach ever-greater speeds. Although Myrabo and Raizer focus on the local problems of launching payloads from Earth into interplanetary space, their concept, like that of a giant laser sail, also has possibilities for interstellar spacecraft that could reach speeds comparable to the speed of light.

Of course, we should not lose sight of the enormous benefits that might arise from the development of laser or microwave sail systems for travel close to Earth. NASA is devoting considerable attention to laser sails as an alternative to chemically fueled rockets capable of traveling through the solar system at low cost. "The vision I have for the future of beamed energy propulsion," Myrabo says, "is to create a technology that could replace airlines [and] spacecraft. Imagine a global aerospace transportation system that runs on solar energy, generated in space and beamed to the vehicle in flight. This kind of transportation system could cut the cost of boosting a payload into orbit by a factor of one thousand below what the Space Shuttle costs right now." The difference between sending material into space at $10 thousand per pound, the Space Shuttle's cost, and $10 per pound would of course be extraordinary. "Imagine a series of space power stations around the planet," Myrabo continues. "Say ten billion watts apiece [far larger than any power station on Earth], beaming power to vehicles in flight. We have the possibility of replacing all the airlines and the Space Shuttle with beamed-energy-propelled vehicles that do not contaminate the planet, are environmentally friendly, run on solar power, and [do not deplete] our fossil fuel reserves. These vehicles can be incredibly efficient, and [can] provide transportation directly to the moon . . . [or to anywhere on Earth] in forty-five minutes. And at the price of a conventional airline ticket today."

Joining Myrabo and Raizer in their enthusiasm for the possibilities of a spacecraft that rides a beam of energy are the engineers Robert Zubrin, the chief proponent of Mars colonization, and Dana Andrews of the Boeing Corporation. Andrews and Zubrin visualize a spacecraft that receives not a beam of light or microwaves but instead one of charged particles created by a nuclear fusion reactor and shot toward the spacecraft, which captures them with "sails" made of superconducting wire, deflecting the particles with magnetic fields to receive most of their energy. Although these particle beams can impart energy to a spaceship far more efficiently than a beam of photons, they cannot be kept as narrow as beams of photons. Hence a tradeoff arises which could leave particle beam propulsion superior within the solar system, and laser or microwave sail systems far better for interplanetary voyages.

If the laser or microwave sail propulsion system should come to fruition for interstellar journeys, a problem would arise when the spacecraft finds that it must decelerate at the far end of its voyage, not to mention additional accelerations and decelerations if anyone aboard plans to head back to Earth. These difficulties might be overcome with more conventional propulsion systems such as those described below; Robert Forward has also envisioned a sail that can be divided into two parts, one of which reflects the laser beam backward onto the rest of the sail, producing a deceleration that slows down the ship for a landing on planets around other stars.

To return to Earth, astronauts would presumably construct a laser acceleration mechanism in a distant planetary system, the mirror image of the lens that gave the spacecraft its initial impulse. Of course, they might instead abandon any thoughts of returning to Earth in person. "We'd have no problem finding volunteers for a suicide mission," says Dana Andrews. "Just think about the things you'd be able to see and name." This sort of reasoning does not apply of course for interstellar journeys that take not decades but centuries. If we send a spaceship on a thousand-year trip to another planetary star system, we would seek and anticipate not so much the return of the intrepid explorers' descendants as the news that they can send to our descendants on Earth via radio and television messages. These transmissions would describe what the explorers find far more efficiently and rapidly than any messenger could.

Given our current life expectancies, those who fund the first interstellar space journeys will not survive to learn of their success even if they watch the launch in their youth. Although we may eventually find ways to make interstellar journeys within a human lifetime (see Chapter 10), these will not be the first. Only with a view of the future far longer than that which we apply to the world today will society devote itself to creating interstellar spaceships. This dream is one we can all admire, though we may disagree on the amount we should contribute to make it come true. Stephanie Leifer, who works on futuristic means of space travel at the Jet Propulsion Laboratory in Pasadena, California, captures this spirit well when she notes that "[interstellar spaceflight] is something that people can justify spending their life's work on even though they know they might not see the results in their lifetime; they can see steps forward to ultimately accomplishing that goal. I don't think there is a single person [who studies the possibilities] who wouldn't love to see an interstellar probe launched, even if they knew they would never see the results come back. . . . I think that what motivates these people is something that starts maybe in childhood; it is something very visceral and deep and in the human spirit, to learn and explore and venture out into the universe as far as we can."

These estimable thoughts leave us with only technological issues to overcome. Having examined the proposed propulsion systems that keep the energy source on Earth, we can survey different types of potential rockets that carry their own motive power, conveniently dividing them into three categories: chemical, nuclear, and matter-antimatter engines. Although these categories do not exhaust the possibilities that our finest minds have hypothesized, they do cover the range of space well because they extend from "mere" stunning technological improvements in current engines to the theoretical best that we can ever hope to achieve.

Chemical Propulsion Systems

Gunpowder rockets, Robert Goddard's first engines, the intercontinental ballistic missiles that military establishments deploy, and the space vehicles that have launched all our astronauts and interplanetary probes—all rely on chemical motors that produce kinetic energy by burning some type of fuel. This combustion is a chemical reaction in which oxygen combines with the rocket fuel to produce new molecular compounds. The new compounds are not quite so tightly bound as the old ones: The reaction converts into kinetic energy some of the energy that was stored in the chemical bonds that held molecules together.

Engineers have devoted many lifetimes to managing the details of which fuel works best and how to manage the combustion process to yield the maximum result from a given amount of fuel. They know full well, however, that all chemical reactions can, even in theory, yield only a maximum amount of kinetic energy, namely the total energy stored in the fuel's chemical bonds before its combustion. No matter what fuel we choose, that maximum can produce only so much kinetic energy to the rocket engine. By providing an enormous amount of fuel, we can create a powerful motor, as we did for the mighty *Saturn V* that sent astronauts to the moon. Nevertheless, the relatively limited amount of energy stored in chemical bonds implies that chemical engines can never take us to speeds greater than a few dozen miles per second. Although these speeds are completely adequate for escaping from the Earth's gravitational domination and can take us anywhere in the solar system in relatively few years, they appear rather hopeless when we dream of interstellar spaceflight. For journeys that cover tens or hundreds of thousands of times our travels within the solar system, chemical rocket engines leave us at velocities much too low for a speedy trip unless we welcome the prospect of spending hundreds of thousands, or even millions, of years on these voyages.

As we have seen, chemical rocket engines can be combined, one atop another, to gain additional velocity. Exactly this arrangement allowed astronauts to reach the moon and also sends automated probes to other planets. However, since each succeeding stage in this arrangement must be smaller than the preceding one, this method quickly exhausts itself long before we can begin to reach the velocities that might make interstellar travel feasible.

Chemical rocket engines seem likely to remain the vehicle of choice for the near future because so much attention has been devoted to the technology of chemical motors that they remain a giant step ahead of other propulsion systems. Eventually, though—and perhaps well before the next century draws to a close—newer, more efficient propulsion systems may supplant chemical engines.

Within the solar system, these new systems include magnetically driven launchers and ramjet engines as described in Chapter 3, which accelerate payloads to enormous velocities in a short distance, hurling them from the Earth's surface into interplanetary space. These mechanisms also rely on chemical fuel, either to produce the carefully controlled explosion of a ramjet, in which fuel steadily ignites behind the accelerating payload, or to generate intense magnetic fields, which then fling matter into space by electromagnetic forces. Like spacecraft that can catch laser beams and microwaves in their giant sails, ramjets and magnetic launchers would keep the driving engine firmly fixed to a planet's surface. Unlike the slowly accelerating push from photons, however, the acceleration produced by these ramjets and magnetic launchers would reach tens of thousands of G forces, devastating to humans but quite useful in launching a series of hard cargo, including fuel, liquid, and other bulk materials, to the far reaches of the solar system at relatively low cost.

To reach the stars, we may have to develop rocket motors that employ something other than chemical reactions. Before we abandon the familiar concept of chemically powered flight, however, we should consider the fact that interstellar space has plenty of potential fuel in the form of the gas (mostly hydrogen and helium) that floats among the stars. On the average, interstellar space contains about one atom or molecule in every cubic inch, with dust grains much more sparsely distributed. This means that an average cubic mile of interstellar space contains about the same number of atoms and molecules we breathe in with every expansion of our lungs. Since a lungful of air hardly ranks as a great supply of fuel for a spaceship, we will have some trouble finding fuel in interstellar space if we merely scoop up matter by the cubic mile. Instead, we must do far more.

Four decades ago, Robert Bussard proposed a rocket that would gather interstellar matter, deflecting it toward its motor not with a physical scoop but with magnetic fields. Much of the hydrogen and other matter between the stars has become positively charged from the effects on it of starlight in the interstellar medium; this means that we can exert forces on it with intense magnetic fields that could guide the trajectories of individual particles as we wish.

The Bussard magnetic scoop might be used to secure fuel for a "conventional" chemical reaction motor. Because oxygen ranks as the third most abundant element in interstellar space, and because hydrogen and oxygen combine to yield large amounts of energy with no more dangerous by-product than water, we can imagine a Bussard spaceship that scoops up interstellar matter and processes this material to separate the hydrogen and oxygen atoms, which are then allowed to combine in continuous combustion. If this seems rather tame, we should turn our attention to still more futuristic designs for interstellar rocket engines, those that use nuclear fusion, not chemical fuel, to accelerate their payloads to speeds close to the speed of light.

Nuclear Fusion Propulsion Systems

We have already seen the importance of the multistage rocket concept, by which each later stage of a spacecraft starts with the velocity imparted by the previous stages. Significant though this may be for chemical reaction motors, the multistage concept finds its most favorable application when we consider interstellar journeys made possible through nuclear fusion explosions. The notion of propelling spacecraft with hydrogen bombs dates back several decades to seminal work by the imaginative physicist Freeman Dyson and his colleagues and envisions a spacecraft with a large "pusher plate" at its rear. The plate protects the spaceship's crew while it absorbs the force from hydrogen bombs exploded, one after the other, at a proper distance behind the spaceship. Since each explosion gives the spaceship a greater velocity, their cumulative effect can push the craft to velocities hundreds of times greater than any chemical rocket system can achieve. Here we have a spacecraft with as many stages as we like, with each stage a nuclear explosion identical to the others in the series.

Why and how could a nuclear-powered spaceship succeed where chemical propulsion fails? The difference arises from the far greater amounts of kinetic energy released from a given amount of fuel by a nuclear explosion, in comparison to a chemical reaction such as combustion with oxygen. As Albert Einstein famously demonstrated in 1905, within any amount of mass, abbreviated as m, there resides a corresponding amount of energy, equal to that mass times the square of the speed of light: $E = mc^2$. The fact that the speed of light is a large number implies that E, the energy of mass, is also large, even for relatively small amounts of mass. The energy of mass that resides within your wallet, for example, could power the United States for a day—if you had a way to convert all its energy of mass into useful forms of kinetic energy. Certainly this beats handing over your wallet to the power company for their far more inefficient production of kinetic energy, which usually involves the combustion of coal or oil.

Chemical reactions do not touch the energy of mass within the atoms and molecules that undergo the reactions; instead, they affect only the much smaller amounts of energy involved in the chemical bonds that hold the atoms and molecules together. In contrast, nuclear reactions involve the nuclei that lie at the centers of atoms and do change the total energy of mass. In a hydrogen bomb, relatively simple nuclei fuse or join together to form other types of nuclei. This fusion converts about 1 percent of the total energy of mass in the fusing particles into kinetic energy—energy that appears as an enormous burst of light and other forms of radiation from the site of the fusion.

One percent may not sound like much, but this conversion of energy of mass into kinetic energy allows a hydrogen bomb to generate about a thousand times more kinetic energy from a given mass of "fuel" than any chemical reaction can. In other words, pound for pound, a hydrogen bomb has about a thousand times the efficiency of the finest motor we have created. This huge advantage has driven scientists and engineers for four decades in attempts to produce a "controlled fusion reactor," a site where nuclear fusion can occur without blowing the machine apart. Such reactors continue to hold great promise for the future because readily available fuel (for example, the hydrogen nuclei in water) could be converted into enormous amounts of kinetic energy. The future will be the rosier for the fact that fusion does not yield the harmful waste products associated with nuclear fission, the decay of radioactive nuclei. Fission, inherently less inefficient than fusion, can and has been used to produce kinetic energy both in "atomic bombs" (more correctly called fission bombs) and in nuclear (better called fission) power plants. In other words, the fact that we can control fission has led to its use as a stopgap until the day when we can use fusion as a cheap, safe, pollution-free source of power.

We cannot say how far we must emerge into the future before the advent of fusion power on Earth, but in space we could begin right away to make fusion-powered spaceships capable of attaining a fair fraction of the speed of light. Dyson and his colleagues have analyzed the physics of accelerating spacecraft in this matter and conclude that we could rather easily create hydrogen bombs of the right size and energy yield to power an interstellar spaceship. Far more difficult would be the creation of a massive pusher plate that would simultaneously protect the crew and, with a suitable shock-absorbing system, receive the blast from detonations behind it. These problems, however, might soon be overcome, and we would have a spaceship fit for interstellar travel.

In a more conventional use of nuclear fusion power, we can imagine a spaceship in which magnetic fields direct the high-velocity particles resulting from nuclear fusion into a particular direction, creating a "magnetic nozzle" that simulates the effect of a more conventional propulsion system. As the history of controlled nuclear fusion on Earth demonstrates, however, this plan is easier to describe than to execute. The best nuclear reaction for a spaceship with a magnetic nozzle would be the fusion of deuterium nuclei with nuclei of helium 3, which we considered mining from the lunar surface in Chapter 3. If such spaceships come to fruition, we may see the time when we sacrifice our satellite to travel to the stars.

The Joys of Antimatter

Of all imaginable spaceship fuels, by far the best is antimatter. Earlier generations of physicists, working at a time when the concept was new and the thing itself never seen on Earth, hyphenated this word as "anti-matter"; today, with antimatter a familiar plaything of high-energy physics, we can drop the hyphen and admire the facts.

"Antimatter" is a catch-all word used to describe the opposite of what we call matter. Like matter, antimatter consists of elementary particles, some of which can join together to form nuclei; nuclei, surrounded by a cloud of lighter elementary particles, produce atoms; and atoms, linking together in pairs or higher-multiple assemblages, form molecules. The difference is that the nuclei, atoms, and molecules of antimatter consist of antiprotons, antineutrons, and antielectrons. Each of these antiparticles has the same mass, but an electrical charge exactly opposite to that of its familiar partner, so that an antiproton carries one unit of negative charge, an antielectron (also called a positron) has one unit of positive charge, and an antineutron has zero electric charge (since the opposite of zero is zero).

Antiparticles can build assemblages of what we chauvinistically call antimatter as readily as ordinary particles combine to form chunks of matter. Antimatter particles experience the same types of forces and undergo the same reactions, such as nuclear fusion, that ordinary particles do. Thus we can easily imagine that somewhere in the universe stars made of antimatter shine just as steadily as stars made of matter do in our own Milky Way. We cannot hope to recognize these stars as "anti" by their light, because an antimatter star produces exactly the same sort of light and other forms of electromagnetic radiation as an "ordinary" star does. Photons, the elementary particles that form the radiation, are identical to antiphotons, and so we call them all photons pure and simple. No amount of detailed analysis of the light from a different galaxy can tell us whether it is made of matter or of antimatter.

One diagnostic for antimatter is failproof: the proximity test. When a particle meets its antiparticle partner, the two particles disappear. In their place, we find only photons and two other types of elementary particles, neutrinos and antineutrinos. What has happened in this annihilation is that the two particles have turned all their energy of mass, the $E = mc^2$ that Einstein made famous, into kinetic energy—the energy of motion carried by photons, neutrinos, and antineutrinos.

In this complete conversion of matter into energy (more precisely, of energy of mass into kinetic energy of particles with near-zero mass) lies the secret for the most efficient fuel in the universe. A spaceship made of matter that carries a good supply of antimatter in its fuel tank is in a position to bring together matter and antimatter in a controlled way, producing bursts of energy that can propel the spacecraft to high velocity.

A little antimatter can go a long way because the energy of mass stored in even a modest mass is enormous. Since the conversion of this energy of mass into kinetic energy occurs with an efficiency of 100 percent, a matter-antimatter spaceship engine has an efficiency about a hundred times greater than one powered by nuclear fusion explosions, which convert about 1 percent of the initial energy of mass into kinetic energy. Since 100-percent efficiency amounts to the theoretical best we can hope to achieve, a matter-antimatter rocket engine represents the finest motor we can hope to create. All we need to do is to assemble matter and an equal amount of its corresponding antimatter for fuel, and arrange for matter and antimatter to meet and annihilate as we choose, producing a stream of high-energy particles to be directed in the direction opposite our proposed line of travel.

Some serious problems arise in the process of making this vision a reality. One is a problem with the spacecraft's fuel tank. If it is made of matter, then far from holding the antimatter fuel, it will explode when the first antimatter touches its surface and annihilates with the matter. No amount of conventional structural strength can deal with this problem. But there is a way out. The solution is to make the fuel tank not of matter (nor of antimatter, which would simply create the same problem in mounting the fuel tank on the spaceship) but of magnetic fields. Because magnetic fields exert electromagnetic forces on electrically charged particles, we can easily (in principle) confine the antimatter and arrange for only a small amount at a time to pass in a specified direction toward the equal amount of matter (also ranking as fuel) with which it will annihilate.

A second problem arises from the fact that the kinetic energy released when matter and antimatter annihilate tends to spew out in all directions and cannot easily be channeled into a particular direction. This calls for the same solution used by nuclear fusion spaceships: a pusher plate that receives the blast from the controlled annihilation events involving predetermined amounts of matter and antimatter. Each annihilation event blasts the pusher plate, and thus the spacecraft, with energy that accelerates the ship to a higher velocity. Gerald Smith, a particle physicist and one of the leading proponents of matter-antimatter spaceship propulsion, happily foresees the day when these spacecraft carry humans to other stars. "All the physics is understood," Smith notes. "The only thing that could go wrong is if we don't get enough [antimatter fuel in a manageable form]. But that's just technology."

Smith is a man after my own heart. I too believe that once we understand the principles of physics, all else is "just technology." Engineers and engineering physicists such as Zubrin, Andrews, and Myrabo lie across a great divide, aware that "just technology" discriminates 100-percent efficiency amounts to the theoretical brates us from the era of hypersonic aircraft capable of traveling at fifteen times the speed of sound and carrying us from New York to Sydney in an hour. "Just technology" keeps us from most of the futuristic scenarios we have encountered in this book, and in this chapter in particular. The question is, will we solve a particular "just technology" problem on a timescale of a decade, a century, a millennium, or a million years? Seen in this light, the "just technology" problem corresponds to "just a matter of time." We as individuals will not live to see laser sail spacecraft or nuclear fusion propulsion or matter-antimatter rocket engines. In fact, we will not live to see anyone leave on an interstellar voyage. But our descendants probably will—if the human race endures for a sufficiently long time. How long? Time, the subject of our next chapter, will provide the answer. I certainly cannot, but this does not prevent me from making a prediction: If human civilization endures, the Y3K problem will concern the selection of those who will travel to other planetary systems in the Milky Way.

Chapter 10

When Time Slows Down

When we look beyond the sun's family of planets, moons, comets, and asteroids, contemplating the possibility of exploring neighboring worlds in the Milky Way galaxy, we encounter distances so daunting that they threaten to cloud our minds. Humans typically react to this situation in one of three ways. They ignore the problem, they imagine the facts to be different from what they actually are, or they brace themselves and confront the difficulty head-on.

Contemplating the Vastness of Space

Historically, the first two approaches have predominated. Today most people accept the fact that astronomical objects are fantastically far from Earth, billions upon billions of miles away. What they still cannot and do not accept is that no matter how great the distance at which cosmic objects can be placed mentally, most of the objects in the universe are still much farther away than that. For example, most of the stars in the Milky Way galaxy are millions of times farther from Earth than the outer planets of the solar system, while a host of well-observed galaxies lie millions of times farther from us than these stars do.

This cosmic arrangement, so straightforward and so unacceptable to our intuitive feelings about the universe, bears repeating. Take a moment to be sure that you have a correct image of the cosmic scheme of things: With the sun modeled by a baseball, the Earth becomes a grain of sand fifteen feet away, while Mars, Jupiter, and Pluto have distances of twenty-five, seventy-five, and six hundred feet from the sun, respectively. In this model, the closest stars to the sun lie many hundred *miles* from the solar system, and the Milky Way galaxy is larger than the orbit of the moon around Earth!

The enormous distances between the stars imply emphatically that if we hope to send spacecraft to visit even the sun's closest neighbors, a fast-moving spaceship is a really useful thing to possess. So far as we know, the greatest speed attainable by any spacecraft lies just below the speed of light, which travels at 186 thousand miles per second or about 6 trillion miles per year. The "just below" arises from the limitations, discussed later in this chapter, that Einstein's special theory of relativity impose on physical reality. Covering 6 trillion miles each year, light takes a few years to journey between neighboring stars, a few centuries to pass from a cluster of stars to the closest star cluster, and a few tens of thousands of years to travel from our position in the galactic suburbs to the center of the Milky Way. If we describe any spacecraft's speed as a fraction of the speed of light, then the spaceship's travel times will equal the time that light takes, divided by this fraction. Thus, for instance, a spaceship traveling at 1 percent of the speed of light—1,860 miles per second, or about 7 million miles per hour—can reach the closest stars in a few centuries, the nearest star clusters in tens of thousands of years, and the center of the Milky Way in a few million years.

The Obvious Advantage of Rapid Spaceflight

These numbers make for long journeys to other stars and imply that interstellar spaceflight is best attempted with spacecraft that can travel at speeds close to the speed of light. In the previous chapter, we examined spacecraft that at least in theory could achieve this goal—a difficult survey involving momentum transfer, antimatter, and hypothetical engineering. We can now reward ourselves by considering the full set of advantages that will accrue when and if humans succeed, by overcoming all the engineering difficulties and dangers, in creating spacecraft that can travel as close to the velocity of light as our descendants may choose.

First of all, these spacecraft would provide the most straightforward advantage of faster travel: We arrive where we are going more quickly and return home (if such is our plan) sooner than we would otherwise. A spacecraft that can travel at nearly the speed of light—for example, at 99.9 percent of light speed—covers distances nearly ten times more rapidly than one that courses through space at "only" 10 percent the speed of light, and ten thousand times more rapidly than the fastest spacecraft built by humans during the twentieth century. This improvement alone suggests that when we contemplate the far future of space travel, our minds should turn to travel at speeds close to the speed of light. But there is more—far more—to add to the advantages of travel at near-light velocities: the benefits of time dilation

When Time Slows Down: The Hidden Advantage of Rapid Spaceflight

The most significant hidden advantage, the one that makes the ability to travel at speeds close to the speed of light so crucial in making plans for future exploration of the Milky Way, lies in the workings of Einstein's theory of special relativity. This theory, first created in Einstein's mind as he worked at his day job (as a patent examiner in the Swiss capital, Bern), implies that if an observer sees a system—any system, from a swarm of bees to a thousand-person spaceship—pass by at nearly the speed of light, he will record that *time passes more slowly in the moving system.* This time dilation, the slowing down of time in a system moving at nearly the speed of light, is not a bagatelle, something that can be explained as a trick of perception, a misunderstanding of how time is measured, or a definitional question of what constitutes a clock or a watch. It is real: The observer will note that everything that time measures, from the aging process of a person or a mayfly to the tick-tick-tock of a grandfather clock, proceeds more slowly in the moving system than it does in the stationary system in which the observer resides.

How much more slowly? The effect remains tiny as long as the velocity involved is less than about one-tenth the speed of light. At lesser speeds (still enormous by human standards), the difference between the rates at which time passes in the moving and stationary systems does not amount to even 1 percent. In other words, unless we are dealing with velocities that lie in roughly the same ballpark as the speed of light, the Einstein time dilation effect can be ignored in all but the most exact applications of physics and engineering. But as we pass 10 percent of the speed of light and head for even greater velocities, the effect grows and grows, technically without limit, as we approach closer to the speed of light. For the mathematically minded, we may present the actual formula:

> The rate at which time passes in the moving system, in comparison to the rate at which it passes for a stationary observer, varies in proportion to the square root of the quantity $1 - (v^2/c^2)$. Here v is the velocity of the moving system, as measured by the observer, and c is the speed of light. If, for example, $v = 99.5$ percent of the speed of light, then v/c will be 0.995 and v^2/c^2 will be close to 0.99. In that case, the quantity $1 - (v^2/c^2)$ will be approximately 0.01, whose square root equals 0.1. In nonalgebraic language, time unfolds in the moving system at only one-tenth the rate that it does in the system where the observer resides. In a more extreme example, if $v/c = 0.99995$ (the velocity v equals 99.995 percent of the speed of light), then (v^2/c^2) will be approximately 0.9999 and the quantity $1 - (v^2/c^2)$ will equal about 0.0001, with a square root of 0.01. Then time in the moving system will unfold at only one hundredth of the rate at which it passes in the stationary system.

For those who do not regard numbers as our little friends, we mention some key benchmarks. At a velocity equal to 60 percent of the speed of light, time in the moving system will proceed four-fifths as rapidly as it does in the stationary system, while at a velocity equal to 95 percent of the speed of light, time will run about one-third as rapidly in the moving as in the stationary system. At 99.5 percent of the speed of light, the ratio becomes 10 to 1: Time in the moving system proceeds at only one-tenth the rate at which it unfolds in the stationary observer's frame of reference. At 99.995 percent of the speed of light, this ratio rises to 100 to 1; and for travel at 99.99995 percent of light speed, time slows down by a factor of 1000.

The strings of nines after the decimal point imply that if we can find a technological means of approaching the speed of light as closely as we like, then we can make time run as slowly as we choose! Time will never slow down completely, but with a sufficiently rapid (and "rapid" hardly begins to tell the story here) spaceship, we can make time slow down as much as we like, as long as we don't insist that it slow down all the way to zero.

The advantages this time dilation effect offers to a space traveler are obvious—provided that she has few regrets for the world she left behind. Traveling at 99.99995 percent of the speed of light, a space traveler could complete a journey of a thousand light-years—far enough to take her past many millions of the stars surrounding the sun—while only a single year passes, for her. Back on Earth, this journey would be seen as requiring just over a thousand years (the "just over" arises from the fact that the travel velocity must always be a bit less than the speed of light), and a round-trip to a star system one thousand light-years from the solar system would take a bit more than two thousand years. At current rates of human aging, this means that during the journey many dozen generations would pass on Earth, even though our space traveler would need only two years, plus whatever time she spends at the far end of her voyage, to complete her explorations.

The Twin Paradox: Isn't All Motion Relative?

Before pursuing this matter further and accepting the notion that Einstein's time dilation effect may someday allow spacefarers to pass from one side of the Milky Way to the other (100 thousand light-years) without devoting much of their lives to the project, we must confront one glaring, eminently apparent problem with the situations we have described above. If an astronaut embarks on a thousand-light-year journey at nearly the speed of light, while her twin remains behind on Earth, won't *each* twin see the other twin as aging more slowly? When the traveling twin returns to Earth, each twin cannot be older than the other! How can we resolve this twin paradox without violating simple logic?

Einstein's theory of special relativity provides the answers to these questions. The theory rests on the postulate that no place and no thing in the cosmos can be identified as truly "at rest" while other objects are in motion.* Instead, as the word "relativity" implies, all motion must be judged with respect to some chosen set of objects, which are denoted as "stationary" for reference purposes. These stationary objects may themselves be considered as moving with respect to other objects. Now doesn't this "democracy of motion" invalidate the assertion that the traveling astronaut ages far less than her twin who remains stationary on Earth?

No, it does not. The democracy of motion described by Einstein's theory extends only to reference systems con-

---

*When Einstein visited Hollywood during the 1930s, one of the Warner brothers supposedly said to him, "I also have a theory of relatives: I don't hire them"—proof that the Warners never fully understood the "twin paradox."

sisting of objects that move in a straight line at constant velocity. For example, if two spacecraft moving in straight-line trajectories pass by one another in opposite directions, each spacecraft has a right to regard itself as stationary and the other spacecraft as moving. In this case, once the spacecraft have had their close encounter, they will never meet again for a comparison of who has aged more rapidly. If one of the two spacecraft changes its direction of motion, the situation will become more complex, and the simple rule of democratic motion will no longer apply. Notice that when different observers (themselves moving with a constant speed and direction) observe a change in the speed or direction of an object's motion, they agree that a change has occurred. This contrasts with the measurement of straight-line motion, which may be observed to be zero if the observer's motion is identical to that of the system being observed.

When physicists apply the theory of special relativity in order to analyze what happens when a spacecraft travels outward, eventually reverses direction, and returns to Earth, they find that the equations imply what we have claimed: The traveling twin indeed returns much younger than the stay-at-home twin. Before citing experimental evidence to confirm this assertion, let us apply our minds to dealing with it, imagining a situation that emphasizes the difference in the experiences of the two twins.

Picture twins, one of whom remains on Earth while the other journeys many light-years into space at nearly the speed of light and then reverses direction and heads home. Suppose that each twin wears a beacon that emits flashes of light once every second. Through impressive technology, each twin keeps an eye out to observe the flashes from the other twin's beacon. The rate at which each twin receives the other twin's flashes indicates how rapidly time is passing for the other twin: Flashes seen at a high rate imply that time is passing rapidly, while flashes received at a low rate imply the converse, that time is passing slowly for the other twin.

The first part of the journey proceeds easily into a straightforward mystery. While the traveling twin rockets away on her outbound journey, each twin sees flashes arriving, not once per second but at a much lower rate. This occurs for two separate reasons. First, time runs more slowly in the moving system, as Einstein's theory states. Second, the distance between the twins is continuously increasing, with the result that each successive flash must travel a longer path before reaching its target. Both of these effects lengthen the time interval between successive flashes received. One of these effects arises from Einstein's time dilation, while the other occurs simply, and without any reference to relativity theory, because of the steady increase in distance. To try out some numbers describing a representative situation, note that if the traveling twin moves away from Earth at 99.5 percent of the speed of light, she will receive flashes from home not once per second, the rate at which the stay-at-home twin produces them, but about once every twenty seconds. Half of the total effect arises from the increasing distance, and the other half from the slowing down of time. The traveler's twin on Earth will likewise receive flashes from space not once per second but instead once every twenty seconds, for the same two reasons described just above.

So far, the twin paradox seems alive and well, since each twin sees the other aging more slowly even after allowing for the second effect, the simple distance increase. But now comes the change that will help to resolve the issue if we pay close attention to what happens when the traveling twin reverses her course and heads homeward. The crucial distinction between the two twins is this: For the traveling twin, the situation changes instantaneously, but this is not true for the twin who remains on Earth. Let us take each twin in turn to explore this difference.

As soon as the traveling twin reverses direction, the two effects that had reinforced each other, each increasing the time between flashes, now find themselves in opposition. On the one hand, the traveling twin's distance from Earth steadily decreases. This tends to increase the number of flashes that she receives each minute from her stay-at-home twin because she is now flying toward the source of light, reducing the distances that each successive flash must cover to reach her rather than sailing away from Earth and steadily increasing these distances. On the other hand, the slowing down of time remains the same, provided that the traveling twin approaches Earth as rapidly as she was receding from it. When you work through the mathematics (the "you" here represents a courtesy stand-in for generations of devoted scientists), you find that, on balance, the traveling twin now observes flashes at a higher rate than would occur if she were not in motion with respect to Earth. The reason for this is that the effect arising from the continuous decrease in distance far overshadows the time dilation effect. If the traveling twin approached Earth at 99.5 percent of the speed of light, then her speed would fall only 1 part in 200 below the speed of light. As a result, in the absence of any time dilation effect, the traveling twin would detect two hundred flashes per second from Earth. (If she traveled at half the speed of light, she would receive two flashes per second by steadily reducing the distance that successive flashes must travel to reach her; at 90 percent of light speed, she would observe ten flashes per second, and so on.) However, the second effect, the slowing down of time, implies that time on Earth appears to run at one-tenth the speed in the moving system that it does on the spacecraft. Hence the traveling twin actually receives not two hundred flashes per second but only one-tenth as many, or twenty flashes per second.

And isn't this exactly what the twin on Earth sees for the flashes from the traveling twin? Close, but not exactly. In the subtle difference between the two twins' motions lies the resolution of the twin paradox. The twin on Earth does detect flashes once every twenty seconds during the first part of the journey and receives twenty flashes per second for the remainder. But unlike the traveling twin, who experiences each of these two situations during precisely half of her voyage, the twin on Earth sees flashes at a low rate for nearly her *entire* journey and receives flashes at a high rate only for a brief time, just before the journey ends.

Why does this happen? The difference arises from the fact that the traveling twin makes an impressive change—in this case a complete reversal—in her motion, while the twin on Earth does not. The reversal causes the traveling twin to receive many more flashes per minute from the instant that she changes her spacecraft's travel from outward to inward bound. In contrast, the stay-at-home twin continues to receive flashes at a low rate long after this reversal because the flashes sent on the outward leg of the traveling twin's journey continue to cover the many light-years needed to reach Earth. It is years before the stay-at-home twin begins to receive many flashes per second—the flashes that the traveling twin sends after beginning her homeward journey. Furthermore, this period of a high flash rate from the spacecraft lasts for only a brief time interval because the traveling twin, moving at nearly the speed of light, returns to Earth only a few months after the first flashes from her inbound leg reach our planet!

The net result is that the traveling twin sees flashes at a low rate for half her journey (the outbound leg), and at a high rate for the other (inbound) half, while the stay-at-home twin receives flashes at a low rate for nearly the entire time that her twin is away and observes flashes at a high rate for only a small fraction of the journey's time interval. This lack of symmetry allows the traveling twin to conclude that a long period of time has passed on Earth (since she receives flashes at a high rate for fully half her journey), while the stay-at-home twin concludes that not much time has passed on the spaceship (since she sees flashes at a low rate for nearly all the journey). The simplified arguments contained in the parentheses in the preceding sentence hold up even after we have carefully and mathematically accounted for the effects arising from changes in the distance rather than for those from the slowing down of time.

Experimental Verification of Time Dilation

If an appeal to the authority of "physicists' calculations" leaves you cold and you find the previous description of the traveling twin confusing, consider the best proof of all that time really does slow down in a moving system: experiment. Three types of experiments have verified the time dilation prediction of Einstein's special theory of relativity. First, physicists have put extremely accurate atomic clocks in aircraft and have shown that these clocks run more slowly by a tiny amount (tiny, but nevertheless measurable) when flown at several hundred miles per hour than they do while resting on Earth. Second, other physicists who use particle accelerators to study elementary particles that decay (break down into other types of elementary particles) have demonstrated that these particles decay much more slowly when they are traveling at speeds close to the speed of light, a situation relatively easy to arrange with modern equipment.

Third, and best of all, astronomers have seen time slow down in the cosmos. In recent years, expert observers have made careful measurements of a class of exploding stars called supernova Ia's (SN Ia's). This type of supernova can be identified by its characteristic spectrum, the relative amounts of light of different colors that it produces at different time intervals after its initial explosion. Supernova observers have found that SN Ia's exhibit a remarkable homogeneity: They all grow steadily brighter for about two weeks after they explode and then decline in brightness at a carefully measured rate which is almost identical from one SN Ia to another. But the preceding statement holds true only for relatively "nearby" SN Ia's, those that explode in galaxies within a billion light-years of the Milky Way.

The expansion of the universe implies that all galaxies are receding from one another with speeds that increase in proportion to the distances between them. Therefore, the more distant galaxies are receding from the Milky Way at proportionately greater velocities. When astronomers study SN Ia's in galaxies many billion light-years from us, they are observing galaxies (and any supernovae within them) that are moving away from the Milky Way at velocities that reach 20, 30, or even 50 percent of the speed of light. For these supernovae, identified as SN Ia's by the patterns in their spectra, astronomers observe the familiar rise and fall in brightness, but the pattern spreads over a somewhat longer time span—as we record it on Earth!

In the decline in brightness of distant supernovae, we are observing the slowing down of time that Einstein predicted, affecting the light from objects that are many billion light-years away and are receding from the Milky Way at half the speed of light. If astronomers exist in the galaxies that contain these supernovae and happen to observe a supernova that appears in our own galaxy (the last one was observed on Earth in the year 1604), they too would see "our" supernova's light rise and fall in brightness at a lower rate than would be true for a supernova close to that distant galaxy. In other words, we see their supernova age more slowly because of its rapid motion away from us, and they see our supernova likewise aging more slowly because we are moving rapidly away from them. Is this a problem? No, because we are not planning to reverse the expansion of the universe and bring the Milky Way close to them; and if we did, the situation would lose its symmetry and resemble the traveling and stay-at-home twins.

The Difficulties of Accelerating Objects to Speeds Close to the Speed of Light

As Henry Thoreau perceptively noted, you never gain something without losing something. Just as Einstein's special theory of relativity adds a hidden advantage to the obvious ones of high-speed travel through space, so too does it pose additional obstacles to the actual attainment of velocities close to the speed of light.

These new difficulties arise from the fact that as any object moves at a fair fraction of light speed, its effective mass increases. In this context, "effective mass" essentially describes an object's resistance to being accelerated to still higher velocities. As the velocity increases, this resistance increases by the same mathematical factor that describes how time slows down. The mathematically minded can refer to the formula on page 196 and once again compute the square root of the factor $1 - (v^2/c^2)$, which describes the change in resistance to acceleration. An object moving at 99.5 percent of the speed of light resists acceleration ten times more than the same object moving at a low velocity; at 99.9995 percent of light speed, the resistance increases to one hundred times its original value. At 100 percent of the speed of light, this resistance would be infinite—a good explanation of why physicists expect that no object with mass can ever travel quite as rapidly as light, whose constituent photons have no mass at all.

These numbers mean that any propulsion system designed to push spacecraft to nearly the speed of light has more work to do than we might have naively imagined. We may have equipment for reaching 95 percent the speed of light, so that time runs at only about one-third of the rate on Earth, but we will find that accelerating the spacecraft is about three times more difficult than we would expect if Einstein's theory were not valid. In our speculation, this poses only a modest problem. After all, if we can imagine a means by which spacecraft can travel at half the speed of light, thousands of times more rapidly than today's best vehicles, why not imagine a bit more, and send them as close as we like to the speed of light?

While we pause to admire the fine logic of this suggestion, let us note one more danger that spacefarers will meet when they reach speeds close to the speed of light. The effective mass of interstellar particles also increases, so that any dust particles, for instance, that the spacecraft encounters will seem to have their masses increased by ten, a hundred, or even more times over their masses when met at lower speeds. Hence a spacecraft making its way through the Milky Way at, for example, 99.5 percent of the speed of light, meets particles that not only seem to be bombarding it at near-light velocities but also appear to have masses much greater than experience would suggest. Defending a spacecraft against these bombarding particles requires another stretch of our supple imagination, and sets an additional obstacle to fulfilling the dream of coursing through the galaxy at nearly the speed of light.

Further Difficulties with Interstellar Voyages at Near-Light Velocities

In addition to the physical obstacles to spaceflight at speeds close to the speed of light that we encountered in the previous discussion, we should also confront the psychological problems involved in sending astronauts on a trip that might last a millennium or a million years, measured by the time that passes on Earth. Who will volunteer for such an expedition? And who will pay for this wild ride through the Milky Way if the participants will return so far into the future that human society, or Earth life itself, might disappear while the astronauts are abroad? Even futuristic societies may balk at funding a mission whose results can be appreciated only by distant generations yet unborn—unless they have developed life extension capabilities that allow them to live indefinitely. In that case, of course, the demand for spaceflight at speeds close to the speed of light will decrease dramatically: If you can live for a million years, there is no need to complete your travels within a century or so.

Until the time that indefinite prolongation of human life becomes a reality, astronauts may prove unwilling to embark for the stars, knowing that everyone and everything they leave behind, save the planet itself, will most likely crumble into dust before they return. Journeys through space at near-light velocities will carry a high price, the permanent loss of all contact with the generations who come and go while astronauts travel for only a few years. Volunteers for these voyages seem likely to include those who hope to see better worlds while abandoning this one to its fate.

In the classic film *Planet of the Apes,* four astronauts ride a spaceship at velocities close to the speed of light, traveling for only a year or so while two thousand years elapse on Earth. (Strangely enough, even this single year seems to require that the astronauts enter a state of suspended animation, as if the movie-makers well understood the boredom of long-term spaceflight.) A malfunction in the spaceship's autopilot brings the astronauts to the wrong planet, but the slowing down of time faithfully follows Einstein's prediction—a rare instance in which the movies treat physics properly.

Planet of the Apes deals more fully with a familiar question in science fiction: What happens when humans first encounter other forms of intelligent life? This issue, so dear to human hearts, brings together our interest in human spaceflight and our natural curiosity and gregariousness. We devote the next chapter to the quest for extraterrestrial intelligence, emphasizing all the while that we may have a good long wait before we have a chance to see how well our imagination has depicted reality.

Chapter 11

Heavy Traffic

Astronomy and astrophysics have their seemingly dull sides, but these lie concealed from view when the subject turns to contact with other forms of life in the universe. Of all the excitement that an ongoing program of space exploration offers, nothing ranks above the prospect that we may soon find ourselves in touch with neighboring civilizations. But this excitement often conceals a false melding of scientific fact and human intuition, which we shall explore fully in the next few pages.

Most people consider the existence of extraterrestrial life a near certainty because they believe (perhaps a bit more fully than the evidence can support) that the universe contains an enormous number of sites favorable to life. What passes almost unnoticed, however, are the implications imposed by this line of reasoning. The same fact that makes extraterrestrial life so likely—the vastness of space—simultaneously makes finding other forms of life extremely difficult. Let us examine the two halves of this thought, taking the easy one first.

The Cosmic Prevalence of Life

Everything we know about the history of life's origin and development on Earth implies that wherever vaguely Earth-like conditions exist in the cosmos, life should appear within a few hundred million years. Scientists (and others) often debate the meaning of "vaguely Earth-like conditions," as well as the definition of "life." We could add our thoughts to this debate but can move more rapidly toward a reasonable conclusion by adopting a conservative approach, one that restricts the search to planets with a composition and temperature much like those of our own planet, and to forms of life that are roughly similar to Earth's. By imposing these restrictions, we obviously miss many possibilities for life on other planets, not to mention forms of life that might exist within interstellar clouds. Even so, we cannot fail to conclude that large numbers of other planets have life on them. The three hundred billion stars in the Milky Way, many or most of which should have planets, provide such a large number of potential sites for life that we can avoid this conclusion only by making the assumption that life requires something more than the right conditions to originate and evolve.

Ever since Copernicus figuratively moved Earth from its central position in the cosmos, scientists have rightly hesitated to assign a special role to our planet, our planetary system, or our galaxy. Of course, this hesitation does not amount to proof; for example, until a few years ago, scientifically acceptable theories suggested that something special (and not currently understood) might be needed for a star to form with planets around it. In that case, even the vast numbers of stars might not furnish many planets with approximately Earth-like conditions. Since 1995 these theories have received their evidentiary refutation from the discovery of more than a dozen planets in orbit around stars relatively close to the sun. These planets have been found not by direct observation (the planets are far too dim and far too close to their parent stars for this to be possible with even the finest instruments) but rather through careful measurements of the light from their stars. As the planets move in orbit, they pull their stars first in one direction and then in another. Astronomers can detect these stellar motions by observing subtle shifts in the colors of the starlight—shifts that arise from what scientists call the Doppler effect, familiar on Earth as the reason for the changing pitch of sound waves when an ambulance or police cruiser sounds its siren while passing by at high speed. Though the evidence for planets orbiting other stars takes this indirect form of observing not the planet but its star, astronomers consider it convincing. As a result, we now know of more planets beyond the solar system than within it.

At present, the method of finding planets around other stars through the perturbations they produce in their stars' motions can reveal only the most massive planets, those comparable to Jupiter, which has 318 times Earth's mass. The conclusion that the cosmos contains large numbers of planets roughly like Earth therefore rests on an extrapolation from what we see in the solar system. If the universe has formed Jupiters in abundance, it should have also formed plenty of Earths—a conclusion to which most experts subscribe. Within the next decade or two, astronomers should deploy new spaceborne telescopes to verify this conclusion directly, adding the weight of evidence to that of authority.

What we really need and want, however, is to find at least one form of extraterrestrial life. Just as the discovery of planets around other stars confirmed the basic concept that Earth plays no special role as a planet, the first extraterrestrial life that we uncover will verify that life on Earth can claim no unique position in the cosmos. Until we find other forms of life, we must rely on the heritage of Copernicus, and on scientific arguments that life seems likely to emerge whenever conditions are favorable, to conclude that life must be prevalent in the universe. Just one other form of life—tiny microbes beneath the Martian soil or floating in the oceans of Europa, for example—would effectively seal the deal that nature has dealt. Until we find it, we shall have a good, convincing argument without any proof we can call definitive.

Nothing daunted, however, we can conclude, along with the bulk of the "exobiological" establishment, that the cosmos teems with life. Having created a class of exobiologists (currently cruelly deprived of specific subject matter), we must add a new field, exobioevolution, whose practitioners will explore and elucidate the patterns that life follows as it evolves independently in various cosmic locales. For now, both exobiologists and exobioevolutionists, like the rest of us, must look to the history of life on Earth to draw conclusions about life elsewhere in the universe.

What does that history tell us? For most of its 4.5 billion-year lifetime, Earth has had life, but only single-celled, extremely simple life, floating in the oceans of our planet. We should not denigrate these simple forms of life, for in addition to evolving into humans, they also changed Earth in significant ways, most notably by making its atmosphere oxygen-rich, about 2.5 billion years ago. Eventually, at a time close to six hundred million years ago, only about one-seventh of the way back through the history of life on Earth, a sudden burst of evolutionary activity occurred, producing large numbers of multicelled organisms of different species. Within a few hundred million years, some of these organisms came to populate the land surfaces of Earth, which came to be covered with plants that several hundred million years ago evolved into trees, flowers, and the grains we eat. The coming of plants allowed animals to evolve on land as well as in the oceans, so that the age of flowering planets, which began about 150 million years ago, was also the age of dinosaurs, who stood atop the food chain for more than 100 million years.

If the dinosaurs had not all died sixty-five million years ago, during the great extinction that occurred apparently as the result of an asteroid impact at the end of the Cretaceous era, they might have evolved larger brains and continued to rule the world, with specialized species capable of writing and reading books such as this one. As things turned out, the dinosaur extinction cleared ecological niches in which another evolutionary product could flourish, with the result that our shrewlike ancestors evolved into large marsupials and modern, placental mammals. A few million years ago, some of these mammals developed particularly large brains, and within the past few hundred thousand years have used those brains to regulate and to dominate the terrestrial environment.

This survey of life's history on Earth should not lead us to think that earlier stages of life proved a failure, for they still thrive all around us as long-term single-celled survivors. However, when we look for forms of life that can engage in conversation with us, we are naturally drawn to what we consider the apex of evolution and can summarize life's history on Earth as 99.9999 percent pre-civilization and 0.0001 percent—the last minute in a year's time—actual civilization, if we place the time when civilization appeared on Earth at about four thousand years ago. If we use the word "civilization" in a term-of-art sense, to specify the ability to engage in interstellar communication and possible journeys, we have had a civilization on Earth for no more than a few decades or do not yet rank as "civilized."

To summarize, all indications point to the conclusion that life of some sort should be abundant in the Milky Way and that time provides the crucial limiting factor governing the extent to which life has evolved in a particular place. If life on Earth is typical, then life elsewhere will spend its first few billion years as only tiny, primitive organisms. Eventually, however, life will display an amazing efflorescence, filling all the available ecological niches to produce a planet with forms of life so diverse that we on Earth have yet to discover and catalog many, if not most, of them. The final step in our extrapolation is the conclusion that a planet with life will eventually produce a species so technologically advanced that it will, if it avoids rapid self-destruction, explore and colonize the regions around it. Hence a few simple steps in extrapolation lead to the notion that within a few billion years after stars and planets form (and they have formed in great numbers throughout the past 10 billion years), many sites in the universe should possess, if only briefly, a society at least as technologically advanced as ours.

When would other civilizations have appeared, or be about to appear, in the Milky Way? About 12 billion years have elapsed since stars began to form in our galaxy. In 12 billion years, many stars can burn themselves out, exhausting their supplies of nuclear fuel, but many others (in fact, the majority) can last longer than this. If intelligent life appears a few billion years after the planet and its star have formed, civilizations might long since have arisen and decayed on the oldest planets in the Milky Way. On the other hand, the Milky Way contains many stars that are roughly the sun's age, and many more that are significantly younger. We can reasonably conclude that if planets with favorable conditions exist in abundance throughout our galaxy, then our civilization on planet Earth arose long after the first appeared, and long before the last will appear. The Milky Way may contain civilizations billions of years older than ours, as well as the relics of civilizations that have come and gone while the few successful ones endured for these billions of years.

But when we ask how the age of our civilization ranks in comparison to those that have already flourished, the answer seems clear: We are among the youngest. If we define a civilization by its ability to communicate across interstellar distances, Earth has had a civilization only during the past century. Compare this time interval with the billions of years that may describe the oldest civilizations in the Milky Way, and our youth becomes strikingly apparent. The quest for contact with other civilizations seems hardly a search for our long-lost brothers but rather an attempt to attract the attention of our grandparents' grandparents.

The Great Fallacy: Earth Is Not the Center of the Universe

The line of scientific reasoning that has led us to the conclusion that there is abundant life in the Milky Way runs headlong into a human failing, one that arises from the intuition that serves us so well in local matters but fails utterly to correspond to cosmic reality. We all know that Earth is a tiny speck of dust in a vast cosmos, but none of us, including the most venerable astronomers, has incorporated this fact into his or her central, id-like awareness. As children, each of us grudgingly had to confront the belief that a single being's existence (our own) could form the center of the cosmos and worked out a compromise in which we learned to relate to others holding contradictory beliefs. Generations of human experience, imprinted on the way that we develop, have left mature adults with an ability to balance their original, childlike concept with the larger reality that surrounds them.

When it comes to the universe, however, experience has had only a modest influence, and our genetic inheritance none at all, in reconciling us to the vasty deep that surrounds our tiny, fragile planet. We can do the math, we can listen to lectures like those in the previous chapters, we can count stars until they pale into seeming insignificance, but our belief remains firm that we reside on a stable, central, enormous world around which the lights of heaven play in amusing and unimportant ways. Though this belief may seem to be under control and incapable of playing a negative role in our lives, it nevertheless rises undiminished to assert itself whenever we muse on the prospect of other civilizations. Then intuition rules, and we feel sure, no matter what we say aloud, that if other civilizations exist, they will surely pay a visit to Earth.

Think for a moment about the issue of contact with extraterrestrial civilizations, and you will recognize the validity of this assertion. Thanks to intuition, reinforced by countless fictional descriptions and depictions of alien contact, we regard the issue as concerning whether or not aliens have visited Earth and, if so, whether or not they can be persuaded to appear on television rather than tormenting us with their ghostly apparitions, crop circles, and crash landings to be covered up by a hostile, all-powerful government. A moment's reflection will show that in these concerns we fall victim to our intuitive beliefs that alien civilizations should visit Earth despite the enormous, astronomical distances that separate us from even our closest neighbors. Suppose, for example, that the thousand stars closest to the sun contain at least one civilization as advanced as ours. This hypothesis puts the nearest civilization closer to ours than most experts estimate, but even so, the distance between the two amounts to about fifty light-years, or three hundred trillion miles—more than a million times farther than the planet Mars. To imagine that other civilizations, if they exist at these distances, would naturally choose to expend their energy and treasure in order to reach our little world requires an astonishing amount of hubris, quite in tune with our intuitive feelings but in violation of the cold facts about cosmic distances.

Isn't it ridiculous that humans labor under the false conclusions induced by their intuition? Yet before we flay ourselves too severely for our failure to incorporate reality into our psyches, let us pause to note that our intuition may have accidentally reached the correct conclusion. Could it be that the lack of extraterrestrial visitors to Earth proves that we are alone as a civilization in the Milky Way galaxy?

The Fermi Paradox: Where Are They?

By devoting some of their deepest thoughts to these issues, professional speculators on the topic of extraterrestrial civilizations have grown aware that the behavior of an average civilization in the Milky Way does not demonstrate how all of them conduct themselves. Since we do not actually know about the existence, development, or behavioral patterns of any civilization besides our own, looking for exceptions to general patterns might seem to be overkill on the speculation front. On the other hand, we made a sweeping assertion in the preceding paragraph that other civilizations—should they exist—would hardly overexert themselves to visit a nondescript planet called Earth. Justice demands that we modify this remark by noting that some advanced civilizations might not fall in line with our notions of how they ought to behave. They might have so much energy to spare or might last for so many millions or billions of years that making a survey of every star in the Milky Way for (among other things) possible life on its planets seems entirely reasonable and worthwhile. Furthermore, if the Milky Way has produced civilizations in abundance throughout the past few billion years, then a small minority of those civilizations must rank as extraordinary for their longevity and curiosity. Why shouldn't we expect that some of these civilizations, which rank among the longest-lived in our galaxy, have probes that come to

Earth not because we are special but simply because we are here, butterflies in the spacious skies of the cosmos?

This question leads straight to what professionals call the "Fermi paradox," named after the great Italian physicist Enrico Fermi. "Where are they?" Fermi rather scornfully asked a roomful of colleagues nearly fifty years ago as they attempted to convince him of the reasoning we have examined. Fermi meant that we lack reliable evidence of extraterrestrial visitors to Earth and that the absence of this evidence implies that advanced civilizations must be extremely rare if they exist at all. The Fermi paradox refers to what seem to be mutually contradictory suppositions, that civilizations are common throughout the Milky Way yet none of them visit the Earth. To resolve the paradox, we can adopt one (or more) of the following hypotheses:

1. Many advanced extraterrestrial civilizations do exist, and a significant number of them do make inspection visits to and past Earth, but we fail to recognize them either (a) because their advanced technology allows them to escape detection or (b) because established scientists and political leaders have decided to conceal their activity from the public to prevent widespread alarm.
2. Large numbers of extraterrestrial civilizations exist, but either laws of physics that we do not now understand or psychosocial rules of behavior that govern these advanced civilizations prevent them from engaging in large-scale exploration of their surroundings in the Milky Way.
3. Advanced civilizations do not exist in the Milky Way in the enormous numbers implied by the most opti-

mistic assessments of the likelihood that these civilizations will appear, flourish, and endure, and those that do exist are not heavily inclined to make their presence known. This explanation embraces the possibility, for example, that we might be wrong about the early stages and that life itself appears only rarely in our galaxy, or about the later stages of a civilization's development, so that, for example, most civilizations may self-destruct soon after reaching something like our present stage of development. On the other hand, even if the Milky Way contains a goodly number of advanced civilizations, they may regard newcomers not as natural sites of interest and exploration but rather as relatively boring, primitive locales best seen as nature reserves.

Let us examine in turn the modes we have created to explain the Fermi paradox. Explanation 1a, that undetectable, advanced aliens make repeated visits to Earth, cannot be refuted. This makes it highly unsatisfying to our psyches, like an attempt to reconcile creationism and evolution by assuming that the cosmos, its fossil and historical record, and our memories all sprang into existence a moment ago. Nevertheless, we can easily concede, as a matter of logic, that we have only the barest of understanding concerning civilizations far more advanced than ours. We can also concede that such highly advanced civilizations could, if they choose, remain undetected by our present civilization with its comparatively low level of technological capability. If highly advanced civilizations do exist and have developed the "nanotechnology" described in the next chapter, they might be able to send probes to all of the approximately three hundred billion star systems in the Milky Way at relatively low cost, measured in energy terms, especially if they are not concerned about a few million years of travel time. These probes might continually send their observations back to galactic headquarters in ways unknown to us now, providing updates on what happens in far corners of the galaxy.

Possibility 1b, that a massive governmental cover-up continuously conceals the truth of ongoing visits from the public, has an appeal roughly one billion times greater than that of possibility 1a, for it ties into our well-founded belief (first presented to me by my grandmother) that the world is run through a series of rackets. It also offers at least the theoretical chance that dedicated researchers, perhaps renegade government employees who are regularly expected to engage in the cover-up, could someday break the shackles of obfuscation and lead us to our cosmic neighbors. As a life-long astronomer and presenter of science to the public, I encounter moments of resentment when I explain that if such a cover-up exists, I must surely be part of it, and find that my audience refuses to consider my status sufficiently elevated to have even a slim chance of belonging to this elite.

Possibility 2, that advanced civilizations either cannot embark on interstellar journeys or choose not to, must be taken to embrace not only physical voyages but also radio and television communications because we currently lack evidence of both actual visitors and messages sent into space announcing the presence of the broadcasters. This seems highly unlikely, for it requires not simply that most civilizations but all or nearly all of them maintain their cosmic distance from us. Our emotions incline naturally to the opposite viewpoint. Humans seem to enjoy announcing their existence to one another and to the universe at large. If we did not, surely someone by now would have pointed out the dangers of continuously emitting large amounts of electromagnetic radiation through our radio and television broadcasts and the radar systems we use to track objects moving through the atmosphere. The leakage from these efforts spreads outward at the speed of light, forming a sphere of photons whose outer edge now lies about a hundred light-years beyond the solar system in all directions, though its inner regions, which extend to perhaps forty or fifty light-years, contain significantly more radio power than its outer parts. Of course, we can easily go wrong by extrapolating our social outlook onto civilizations far more advanced than ours, including our own in the distant future.

Possibility 3 appeals to most scientists as the most reasonable answer: We are not visited because space travel costs a bundle and because the Milky Way contains an intermediate rather than an enormous number of advanced civilizations. If our galaxy had, for example, ten thousand civilizations significantly more advanced than our own, the average distance between neighboring civilizations would be about a thousand light-years—a region of space containing more than ten million stars. But this brings us back to the difference between the behavior of most of these civilizations—say 9,950 out of the hypothetical ten thousand—and all of them. What holds true for the vast majority may be completely untrue as a description of the totality. Shouldn't at least some of these assumed ten thousand civilizations come by to observe and to catalog us, if not to amuse or consume us?

In coming once again to the Fermi paradox's lively impasse, we have sailed by one facet of possibility 2, the notion that advanced extraterrestrial civilizations exist in large numbers, have long since formed a galactic network of intercommunicating societies, and have decided on cosmoecological grounds to respect the right of developing worlds to proceed without outside interference. This "zoo hypothesis," as it is called, has the defect of being completely unverifiable. Unlike the hero of the movie *The Truman Show,* who sailed his boat to the edge of the set, we have no way to break through the curtains of reserve that hypothetical advanced civilizations have imposed on what otherwise might be attempts at contact.

Or do we? One avenue beckons in our quest to resolve Fermi's paradox: interstellar spaceflight. If the Milky Way contains ten thousand civilizations, with an average separation between neighbors of a few hundred light-years, then we need only travel distances of this magnitude, and visit the few million stars closest to the sun, to find the closest civilizations. If we make these voyages and find no one, we must revise either our estimate of the abundance of advanced civilizations or their desire and ability to conceal themselves from our detection. Could it be, however, that developing the capacity to make these explorations will bring us to the magic moment when the "galactic club" opens its doors to a newly qualified member?

Searching by Radio versus Searching by Spacecraft

If membership depends on the capability of interstellar spaceflight, we have a long wait ahead of us. As discussed in Chapter 9, spaceships capable of interstellar voyages seem to lie a long way in the future—a long way, that is, in human terms, though quite possibly an eye-blink in the history of our planet. If we produce interstellar spacecraft in, say, the year 2599, that will hardly count as a longer time than the year 1999 in the scheme of the cosmos, in which the age of the Milky Way has now reached something like the year 11,856,902,147, and the sun and Earth may have begun year 4,579,246,051. From a strictly personal viewpoint, however, those eager to contact other civilizations never lose sight of the fact that we have already acquired the ability to do so by using radio and television. The search for extraterrestrial intelligence (SETI) efforts in the United States and other countries are in their infancy, but dedicated astronomers pursue them, striving to detect radio signals, either deliberate messages or local broadcasts on whose leakage we can hope to eavesdrop. To date, all SETI efforts have taught us a good deal about the detection of faint signals but have actually found evidence of only one civilization, the radio noise from Earth.

We can hope to improve our SETI efforts mightily during the next few decades by developing more sophisticated equipment that will allow us to listen simultaneously to more radio frequencies (billions of possible frequencies exist on which either beamed or leaked messages could travel) and to detect fainter signals than we can now pluck from the background of our own civilization's noise. However, if all other civilizations conform to the zoo hypothesis, these efforts will prove nugatory, and we will remain alone, so far as we know, in the Milky Way and the mighty cosmos. This is one area in which spaceflight has an advantage over cheaper, cleverer techniques: If we visit other worlds, we will have a better chance of finding their inhabitants, even if they are trying to hide from us. This raises the specter of how those who don't want to be found will react when we find them, but we can safely leave this issue, like many others, to future generations who will consider the possibilities of interstellar journeys that may reveal the truth about extraterrestrial intelligence.

Chapter 12

What Don't We Know?

Our assessment of possible human voyages through the Milky Way has led us to conclude that great improvements in our technological abilities, along with much increased confidence in our long-term survival as a civilization, seem to be mandatory in order for these journeys to become reality. Even then, voyages to other planetary systems must last for decades, centuries, or millennia, as far as time measured on Earth is concerned, so that those who cheer brave explorers on their departure are unlikely to welcome them home. We have reached these conclusions by extrapolating from what we know now as we look into the future as far as we dare—not a serious restriction as far as this book is concerned.

Could it be, however, that rules of physics, completely unknown to us now or only dimly suspected, will yet stand our current knowledge on its ear? Will our descendants leap from star to star as easily as they do on television, looking back with a mixture of scorn and pity at our bumbling, exhausting efforts to colonize even the closest celestial objects? The brief answer to these questions is of course, that only time will tell. A longer answer takes us to the delicious irony that extrapolation from our current knowledge leads to the conclusion that the universe contains far more of what we do *not* know than of what we *do* know.

Fittingly enough, even that conclusion does not pass unchallenged. Some writers at the end of the twentieth century, just as at the end of the previous one, claim that we are now approaching the "end of science," in possession of the broad outlines of all important areas of biology, chemistry, physics, and cosmology, with only the details to be filled in. Those who know the history of science generally feel that to state this verdict is to refute it, since nothing from the past suggests that it could be valid. But there is a first time for everything, and once again only the future can conclusively resolve this issue.

If we ask what conceivable advances might make interstellar voyages much easier than we now believe, three broad areas can supply an answer. One deals with rocket propulsion and raises the possibility that we might find an easy way to accelerate spacecraft to nearly the speed of light, gaining all the advantages described in the previous chapter. The second takes us into the realm of life extension, either for individuals or for human society. If we came to regard a million years in the same light that we now deal with a decade, then a voyage lasting half a million years, for example, would hardly seem formidable. The third approach looks toward a deeper understanding of space that will reveal to us how to find and how to use "cosmic wormholes," shortcuts from one part or the universe to another. A variant approach examines the concept that the speed of light does not in fact provide an ultimate speed limit and that we may find ways to travel much more rapidly than 6 trillion miles per year. Finally, to be complete in our speculation, we ought to include the possibilities that interstellar journeys will cease to be an issue in human society because we eliminate either the society or the desire to embark on such voyages.

The Quest for Faster Rockets

On the rocket propulsion front, we can fairly claim to have exhausted our current knowledge with the survey made in Chapter 10. There we concluded that the most efficient conceivable fuel (conceivable, that is, in light of our current understanding of the physical universe) consists of a matter-antimatter mixture. So long as we can arrange for the actual mixture to occur only when and where we want it, and for the products of matter-antimatter annihilation to accelerate our spacecraft in the proper direction, nothing that we know offers a better way to propel spaceships to the far reaches of the galaxy or even beyond. This conclusion rests on the principle of the conservation of mass energy, a bedrock foundation of the laws of physics as we understand them. We may yet see this rock overturned and a bright future emerge from beneath it, but for now the most penetrating conclusion to be made on the better-fuel issue is that if we find something superior to combining matter with antimatter, it will be a stunning, revolutionary moment in our knowledge of the cosmos.

Life Extension: What Are the Consequences?

In contrast, imagining serious life extension on either an individual or a human society basis seems eminently reasonable, almost likely on days when we do not regard the human prospect as hopeless. Even today we seem almost on the verge of being able to grow new organs for human bodies, capable of replacing those that wear out with time; if this proves feasible, the process could presumably be expanded and repeated so that each of us could become a near-immortal organic version of Frank Baum's Tin Woodman. This may seem a dubious projection from today's situation, when we may be "close" but have yet to grow a single new organ for a human body, but those who ought to know regard such a capability as something close to a sure bet for the next century.

If life prolongation becomes a reality, to the point that each human becomes nearly immortal in terms of our present situation, the implications for human society will be staggering. Old science fiction stories about begging Grandpa to cease his existence in order to let a new child enter the world could approximate actual conditions on Earth unless we find a way to open successive new worlds for colonization. On an even deeper level, propagation of the human species would undergo a sea change once individual human lives could be extended nearly forever. Whether the desire to produce children will survive the news that death is avoidable furnishes a question that many would rather not contemplate; the good news is that our generation at least will not be required to do so.

On the other hand, we may rather easily conclude that a spacefaring society that sees itself as effectively immortal, either because its individuals can avoid death or because it has reached a point of utter confidence in its social survival, will regard interstellar space travel in a light entirely different from that cast on it by a rapidly evolving society of short-lived individuals. One possibility for such "immortals," already raised in the previous chapter, is that they will lose interest in space travel, either as part of the intellectual attitude that arrives with the loss of death or as the result of establishing effective interstellar contact through radio and television.

Nanotechnology and Little Slow Rockets

A long-lived, highly developed civilization seems relatively certain to develop nanotechnology, the art of making things small. Simple extrapolation from our own success in reducing the size of computer chips and related technology suggests that we may see progressively smaller equipment, limited at the low end only by the facts of physics. These accomplishments imply that we may eventually make machinery from individual atoms and molecules, whose sizes are measured in billionths of an inch—the "nano" in "nanotechnology." (To be precise, here scientists are talking about billionths of a centimeter, but we can overlook this distinction as we unleash our predictive powers.) In theory nothing prevents us from constructing nanotechnological spaceships, no larger than a grain of dust, and sending them like dandelion seeds throughout the Milky Way. These miniature explorers might not travel at extremely high speeds, for they would be relatively fragile, but we might also find a way to accelerate them with electromagnetic forces to velocities to the speed of light as close as we choose. Each of these nanoships could send back images from its travels, so that before long—perhaps only a few million years from now—we would have an ongoing cinema of the cosmos, a gallery of the worlds in our galaxy, as well as pictures of the immense clouds of interstellar gas and dust within which new worlds are being born.

If we could do this, so could other civilizations. As mentioned in the previous chapter, we have no way of disproving the notion that nanoships from abroad have long since examined the Earth, along with every other object in the Milky Way, capturing our triumphs and failures and possibly subjecting our planet and its inhabitants to a screen test. Could we have already been blackballed for admission to the galactic club thanks to our poor performance? Or could we stand on the edge of success, with only a few more good moves before the doors are opened? Might it prove true (may the cosmos forfend!) that we have awed the Milky Way with our stunning development, so that other civilizations are nervous about approaching our majesty? Thanks to the rules of cosmic development, the latter possibility can hardly be reality; instead, we must rank among the youngest of galactic civilizations, assuming that others exist.

Cosmic Wormholes

If we hope seriously to investigate these possibilities, we may have to find ways to travel through the galaxy. Even if we never learn how to give individuals immensely long lifetimes, we might learn how to make interstellar journeys far more rapidly than now seems possible by establishing the reality of cosmic wormholes. A cosmic wormhole is a tunnel through space-time, a topological feature in the fabric of space that bends space back on itself with the result that what we call a straight-line journey from point A to point B actually furnishes the long way around.

This concept is easy enough to describe but, like the rest of the speculation in this chapter, a good deal harder in the execution. The possibility of using cosmic wormholes to shorten interstellar travel depends on the successful resolution of at least two separate problems: Do cosmic wormholes exist? And if so, how can we use them for our ends, namely sending explorers from here to there on short, rapid trajectories?

We would do well to keep these two questions separate in our minds because the news from the cosmological frontier seems to be that cosmic wormholes could exist, but for now at least, we have no reasonable way to exploit them for travel purposes. The great name in wormhole studies is that of Kip Thorne, a physics professor at the California Institute of Technology, whose reputation for insight and balance is unchallenged by his peers. Thorne has been considering the possibilities of wormholes for fifteen years, since his friend Carl Sagan, deep in writing his science fiction novel *Contact*, asked him for a scientifically plausible way to travel quickly through the Milky Way. Although he enjoys effective interactions with his physics colleagues, Thorne considers himself a loner, a word that accurately describes his eagerness to engage the public at large. Quite understandably, Thorne hesitated to publish his conclusions about wormholes because he knew that they would eventually provide grist for the tabloid mill, which has its way of reinterpreting even the most arcane articles in *Physical Review Letters*. And indeed this came to pass: Thorne had to deal with some of the most outlandish questions that journalists ever posed to a scientist.

This result seems inevitable when we examine what Thorne has concluded about wormholes. According to Thorne and his coworkers, our present understanding of the physical universe does allow the possibility of short, strange stretches of space-time that connect regions apparently separated by millions of light-years of distance. Nor is that all: Each of these cosmic wormholes, once in existence, would furnish us with a time machine capable of sending travelers on journeys back through time. Though this raises problems with what physicists call causality, and the public refers to as the "kill-your-parents paradox," to Thorne the mathematical logic that raises the possibility of cosmic wormholes and their behavior as time machines seems clear. Cosmic wormholes may be unlikely, possibly nonexistent in the entire cosmos, but Thorne and his associates believe that they are not forbidden by the universal laws of physics as we understand them.

If we allow the possible existence of these wormholes, the next question we might address deals with survival while passing through one. In conversational circles where cosmic wormholes are a familiar topic, one often hears a debate as to whether the passage of a wormhole might be as disruptive as falling into a black hole, a region of space where gravitational forces become incredibly strong, capable of ripping human bodies, or even spaceships, completely apart by tidal forces. Earlier considerations about cosmic wormholes speculated that they might involve a black hole at one end and a "white hole" at the other, where material spews forth as forcefully as it falls into a black hole. Thorne's concept of cosmic wormholes, however, severs this link, for he treats them as separate entities, neither black nor white, just basic wormholes that (at least in theory) could allow us to leap the light-years and travel backward in time.

Concerning the more important practical question of whether cosmic wormholes exist, our current knowledge is silent. Stephen Hawking, Thorne's famous friend and fellow cosmologist, strongly believes that nature does not allow wormholes. He conjectures that as we gain increased knowledge of the laws of physics, we shall perceive that no time machine can exist, a speculation, he points out, that will "keep the world safe for historians." What keeps us from knowing whether Hawking is right, or whether Thorne's conjectures might actually describe some part of the cosmos, is our limited understanding of gravity, the force that bends the fabric of space-time.

Finding a Theory of Quantum Gravity

What we lack is a theory that combines an understanding of gravity with our knowledge of quantum mechanics, the physics of extremely small distances and time intervals. Even now, despite seven decades of effort, physicists cannot find a coherent and successful way to meld Einstein's theory of gravity—the general theory of relativity—with the laws of quantum mechanics. Thorne and Hawking rank high among those who have tried and are trying to do so, but so far this goal still eludes them.

Thorne and Hawking do agree on one aspect of wormholes, however: The cosmos offers no compelling reason for them to exist, even if their existence is mathematically allowed. Unlike black holes, which arise naturally when the cores of burnt-out stars collapse, the production of cosmic wormholes appears (as Thorne sees it) to be complex, careful, directed efforts by a civilization far more technologically advanced than ours. Just this concept appears in Carl Sagan's novel and continues to provide the "scientific" basis for a multitude of movies and television shows, in which our favorite characters can use wormholes to slide from one part of the cosmos to another or alien civilizations can employ them to appear and disappear as if by magic. (It is worth noting that the words "magic" and "machine" have the same root, which also provides the source of "magus," a sorcerer with occult powers; our language reflects a time, such as the present, when most people see little or no difference between technological advances and magic.)

Travel at Speeds Much Faster than Light

As physicists continue their attempts to meld gravity with quantum mechanics, they also speculate about the possibility that the speed of light does not impose the ultimate barrier to rapid travel through space. In Chapter 10 we encountered the limitations imposed by Einstein's special theory of relativity. This theory implies that as we attempt to accelerate any object with mass to progressively greater speeds, its effective mass grows larger, and so we must provide progressively greater amounts of energy in order to achieve even a tiny acceleration; reaching the speed of light itself would require an infinite amount of energy, which no one has.

At least two means exist for surmounting this obstacle. One is to conclude that since Einstein disproved whoever came before him, someone will doubtless eventually disprove Einstein, and we shall find a means of accelerating spacecraft to speeds far beyond the speed of light. This approach has the merit of simplicity, but it contradicts the volumes of evidence that support Einstein's theory, including a perfect match between the theory's predictions and what physicists measure when they shoot particles through giant accelerators. The second way to breach the barrier lies in assuming that we need not accelerate any particles to and beyond light speed. Instead, we merely need to find particles that *always travel more rapidly than light* and therefore avoid the problems we encounter as we rise from below toward the speed of light.

These particles exist—in the imaginations of physicists. Einstein's theory of relativity allows the existence of "tachyons," particles that always travel more rapidly than light, whose name derives from the Greek word for "swift." Tachyons have many remarkable properties (on paper). When we seek to accelerate particles that move more slowly than light, we must add energy to do so, but for tachyons the situation is reversed, and we must add energy in order to slow them down. To slow them down "all the way" to the speed of light requires an infinite amount of energy, the same amount needed to accelerate an "ordinary" particle to this speed.

Without going further into the amazing properties of tachyons, we should note several aspects of their hypothetical existence tending to diminish their usefulness for future space travel. No theoretical avenue appears to allow conversion of a tachyon into an ordinary slower-than-light particle, or vice versa. Hence the interstellar journeys that tachyon matter may make cannot take our familiar matter with it. This might leave us relatively content to talk to the tachyon people (if they exist) who zip from place to place as rapidly as we choose to imagine, living vicariously off their travel sagas. But here too theory interposes a problem: Our present understanding suggests no way for tachyons to interact with ordinary matter. This leaves open the untestable possibility that tachyons continually pass by Earth, and everything else we call matter, profoundly unaffected and incapable of demonstrating even their existence, let alone photographs from their travels.

Perhaps this is just as well. Travel by tachyons seems just too easy, a violation of the principle that unless you work to achieve it, you can't really enjoy it. Humans relish formidable tasks that offer great rewards, and the prospect of interstellar journeys poses as great an achievement in this category as we can easily imagine. Surely future generations—perhaps not of the next millennium but within a few millennia thereafter—will meet this challenge. Or will they?

Will Future Societies Engage in Space Travel?

An essential element in projecting the future of human space travel deals with the human mind, and more particularly with the collective attitudes of what we call our civilization. Humans have achieved world domination by combining an inquisitive disposition with a capacity to manipulate their environment, originally with opposable thumbs and forefingers, and later with bulldozers and dredges. After spreading over Earth's surface by the billions, we have begun to explore the ocean depths and have sent humans to the moon and automated spacecraft to the planets. Of all extrapolations into the future, one of the most straightforward assumes that humans will continue the habits of previous millennia by seeking new worlds to explore and to inhabit.

But since we are probing the limits of our knowledge, we might also ask whether human society will lose the urge to settle new worlds or even to investigate them. Our exploratory attitudes, leading naturally to an expansion of territory and population, have squeezed six billion of us onto planet Earth, with the prospect of double or even triple that number before we are halfway through the next century. This superabundance of humans stretches our resources and leaves at least one-third of the world's population in a state of permanent crisis.

How can this problem be resolved? Space colonization cannot provide the answer, despite the pious hope of some theologians who oppose birth control. Building spacecraft to carry a million emigres into space every day appears to be at least as difficult as developing sites on which these potentially unwilling colonists might live. The undeveloped acreage in the solar system, not counting the four gas-giant planets, which offer no solid surfaces on which to live, amounts to less than four times the total land area on Earth. One-third of this land lies beneath the carbon dioxide clouds of Venus, which trap so much heat that the planet's surface swelters at 800 degrees Fahrenheit; one-third of the remainder lies on Mars, whose terraforming we considered in Chapter 5; and the rest resides primarily on Titan, on Mercury, on Jupiter's four large moons, on Neptune's moon Triton, and on our own moon. Even if we ignored the mammoth engineering difficulties involved in providing these objects with Earth-like conditions, we would only postpone, by less than a century, a situation in which we would find all of them as crowded as Earth is now.

This analysis fully justifies the conventional wisdom that we shall find the solution to Earth's population crisis right here on Earth, presumably through some form of birth control. All studies show that the best means to this end lie in raising the standard of living, with the result that economic and social pressures to have numerous children will decrease. Should we be sufficiently fortunate and intelligent to achieve this result on a global basis, we may be able to reach a situation in which the population remains stable and human society views itself as capable of indefinite survival. However, the ineluctable concomitant of such a steady-state existence might prove to be a lack of interest in enormous projects that carry the possibility of great risks and equally great rewards. A society without life-threatening problems might be one without the exploring bug, and this might be a universal rule. Since we can expect life and civilization to arise on planets, which by the rules of physics have only finite amounts of area, we may well find that other civilizations, like our own, first explode in population as they grow technologically adept and then face the crisis bred by their own success. If they either destroy themselves (as everyone knows we might, given the destabilizing effect of worldwide hunger and fear) or reach a sustainable situation with essentially zero population growth, they too might vanish from the exploring realm, content on their own worlds and uninterested in finding new ones to populate.

This analysis also offers a solution to the Fermi paradox; it explains the absence of extraterrestrial visitors to Earth not by assuming a lack of these civilizations but rather by positing a universal rule that highly developed civilizations stay at home. A more detailed speculation, however, suggests a possible flaw in this line of reasoning. As a civilization acquires spacefaring capacity, it might plant colonies in its neighborhood as its population crisis intensifies, just as humans may. The colonies may then adopt a political and social life of their own, as the United States once did, independent of the motherland. Then, while the home planet reaches a steady-state solution, each of the colonies could in turn expand, establishing its own colonies before its population problems cause it to adopt a "mature," nonexploring attitude. The net result would be that the total volume of space inhabited by an original society and its descendants would continue to expand as world after world finds itself home to a mature society with no urge to explore.

Will Extraterrestrial Civilizations Tell Us of Our Own Future?

One shortcut to the future might manifest itself at any instant—or may never appear in the history of Earth. If we enter into contact with other civilizations, we shall presumably acquire a chance to share in their stock of knowledge garnered from their own experiences in the Milky Way and even beyond. At the price of some humiliation, we may learn of the patterns of growth and decay that civilizations undergo, following laws of cosmic development which we can now only guess at and, with our single example, emulate. But even as we learn of the fate that most likely awaits us, or profit from the failures of other civilizations to see what we must do to preserve our own, we may also revel in hard-won insights into physics and cosmology that now elude us. We might even learn which questions simply cannot be answered. What sets the limits to our knowledge? In the words of the astronomer Sandra Faber, “I think this is a profound question, and I don’t know that we’ll really know the answer until we find other beings who are perhaps smarter and more advanced.”

With this thought we reach an appropriate end to our survey of space exploration, a journey through our solar system, the Milky Way, and the intergalactic spaces beyond. During the twentieth century, we humans have taken the first steps on this voyage. We may reasonably expect our descendants to continue to expand our cosmic horizons, and perhaps before long to meet other civilizations that have embarked on their own journeys. In asking you to let your imagination travel freely through space, I hoped to promote the feeling that we are citizens of the cosmos rather than simply members of the tribe on planet Earth. Let us do what we can to remain proud of our tribe, able to defend its actions in the courts of space—or at least in those of our own consciences.

Further Reading

Brandt, John, and Robert Chapman. *Rendezvous in Space: The Science of Comets*. New York: W. H. Freeman, 1992.

Chandler, David. *Life on Mars*. New York: Dutton, 1979.

Goldsmith, Donald. *The Hunt for Life on Mars*. New York: Dutton/Plume, 1997.

Goldsmith, Donald. *The Ultimate Planets Book*. New York: Quality Paperback Club/Byron Preiss, 1998.

Goldsmith, Donald. *Worlds Unnumbered: The Search for Extrasolar Planets*. Sausalito, CA: University Science Books, 1997.

Lewis, John S. *Mining the Sky*. Reading, MA: Helix Books, 1996.

Ley, Willy. *Rockets, Missiles, and Space Travel*. New York: Viking Press, 1951.

McDougall, Walter A. *The Heavens and the Earth*. New York: Basic Books, 1985.

Mermin, David. *Space and Time in Special Relativity*. Ithaca: Cornell University Press, 1968.

Morrison, David, and Tobias Owen. *The Planetary System, 2d. ed.* Reading, MA: Addison-Wesley, 1996.

Norton, O. Richard. *Rocks from Space*. Missoula, MT: Mountain Press, 1994.

Ordway, Frederick, and Mitchell Sharpe. *The Rocket Team*. New York: Thomas Crowell, 1979.

Sagan, Carl. *Pale Blue Dot: A Vision of the Human Future in Space*. New York: Random House, 1994.

Shostak, Seth. *Sharing the Universe: Perspectives on Extraterrestrial Life*. Berkeley, CA: Berkeley Hills Books, 1998.

Thorne, Kip. *Black Holes and Time Warps: Einstein's Outrageous Legacy.* New York: Norton, 1994.

Zubrin, Robert. *The Case for Mars*. New York: Simon & Schuster, 1996.

Index

C

D

E

G

H

I

T

About the Author

Donald Goldsmith is the author of *The Ultimate Einstein, The Hunt for Life on Mars,* and *The Astronomers.* He graduated from Harvard and received a Ph.D. in astronomy from the University of California at Berkeley. He has won numerous awards for his writing from various scientific associations, including the prestigious American Institute of Physics' Science Writing Award and the American Astronomical Society's Annenberg Foundation Award for education in astronomy. He was a consultant for Carl Sagan's "Cosmos," and has served as science editor and co-writer on several science programs and series for PBS. He has also served as visiting professor/lecturer at Stanford, the University of California at Santa Cruz, Berkeley, and Irvine, and Cornell. He lives in Berkeley, California.